Nature not Mocked

PLACES, PEOPLE AND SCIENCE

Nature not Mocked

Places, People and Science

Peter Day
The Royal Institution of Great Britain, UK

ICP
Imperial College Press

Published by

Imperial College Press
57 Shelton Street
Covent Garden
London WC2H 9HE

Distributed by

World Scientific Publishing Co. Pte. Ltd.
5 Toh Tuck Link, Singapore 596224
USA office: 27 Warren Street, Suite 401-402, Hackensack, NJ 07601
UK office: 57 Shelton Street, Covent Garden, London WC2H 9HE

British Library Cataloguing-in-Publication Data
A catalogue record for this book is available from the British Library.

NATURE NOT MOCKED
Places, People and Science

ISBN 1-86094-576-7

Typeset by Stallion Press
Email: enquiries@stallionpress.com

Printed in Singapore by World Scientific Printers (S) Pte Ltd

Contents

Preface

We often forget that the science underpinning our contemporary civilisation is not a marmoreal edifice, fixed forever in its present shape. On the contrary, at each moment as it developed over past centuries, it grew and changed by the efforts of individual people and the institutions they created. Therefore, the tapestry of disciplines that we call by the generic name 'natural science' does not only consist of facts uncovered about the world around us and the laws that connect them. As arguably the finest single product of the human mind, its substance and direction have been strongly conditioned (some might even say determined) by the people drawn to take part in the enterprise and both the physical and social environments in which they have worked. Having had the good fortune to be associated with numerous scientific institutions in various countries over the last forty years, I have had the chance to observe how they came to be what they are, as well as getting acquainted with some of the remarkable personalities (past and present) whose lives and characters have shaped them. In particular, as Director of the Royal Institution of Great Britain and its Davy Faraday Research Laboratory for most of the 1990s, I came to see how that unique body grew out of the preoccupations and personalities of its founding fathers, evolving continuously to meet the challenges of successive generations.

As a result of that background, and in particular the part played by the Royal Institution in what has rather pompously been called 'public understanding of science', from time to time I have written articles and essays on people and places connected with science, as well as the favourite topics that I have worked on myself. From the beginning, I was keenly aware of the social and historical context in which present-day science must be placed, and that provides the thread linking the topics collected here. Broadly, they divide into three categories: places, people and science. Pride of place in the first category goes to the Royal Institution, but with a side-long glance at others, especially an international laboratory in France, the Institut Laue-Langevin. The second category, people, is divided between giants of the past and some present-day practitioners whose lives I find especially remarkable. As to the third—science—I have plundered the texts of Friday Evening Discourses that I gave at the Royal Institution, as well as other popular accounts of research areas that are still developing and to which I have been able to contribute.

My thanks are due to the owners of the original copyrights on the articles reproduced here. I have edited them to a certain degree to take account of more recent happenings but inevitably (and perhaps it may even be a source of interest) they betray their origins in the times when they were written.

Peter Day
Oxford, January 2005

part 1

TEMPLES OF SCIENCE

Science, as a tool for understanding the natural world and ultimately controlling it, is arguably the greatest single adventure of the human intellect, and across the planet, it goes on in an astonishing variety of organisations and premises. They may be local, regional, national or international; university departments or research institutes; huge pieces of kit the size of an automobile assembly plant or small rooms full of flasks and beakers, according to those aspects of the natural world that give them their focus. But not only that: each one is also the result of individual efforts by concerned groups of people who decided, at a particular moment in time, to set up something new in a particular place. That, as well as the exigencies of intellectual enquiry or national need, should never be forgotten. In such a spirit, these opening pages concentrate on two organisations of very different character in two different countries. In fact, the only feature they have in common (and which gives me the excuse for writing about them) is that I spent several years in each and came to know and respect them.

Of all the organisations promoting science that are known to me, easily the most unusual is the Royal Institution, to be found at Number 21 Albemarle Street in the middle of Mayfair in London. In fact, it is unique from several points of view: its peculiar status as a kind of club, independent of government; its longevity (205 years as I write); the way it combines research with outreach of science to the community and, finally, the astonishing number of profoundly significant discoveries that have been made there. How it came into being, survived, flourished and adapted is a story well worth dwelling on.

My second example (the Institut Laue-Langevin in Grenoble) is altogether bigger in scale (a staff of nearly 500 and an annual budget of £33 M at the time I was its Director 15 years ago); a multinational endeavour in southeastern France housing a nuclear research reactor and some 30 large instruments. But that, too, is a result, not just of scientific priorities prevailing at the time it was set up (much more recently than the Royal Institution, of course: in fact, in the 1950s), but also of politics and personalities.

Finally, in this part, I want to draw attention to the family shrine at the heart of most 'temples' of my title: the coffee room. If you ask how ideas (and not only about science) get promulgated, shared, criticised and validated first, before they reach the wider world, then look no further.

chapter 1

The Royal Institution: Then and Now

The Beginnings

In 1999, the Royal Institution (RI) celebrated its bicentenary. The formal decision to found this remarkable organisation can be traced to a meeting that took place on 7 March 1799 at the house of Sir Joseph Banks at 32 Soho Square in London. Sadly, the house was demolished in the 1930s to make way for an office block, but a splendid souvenir of it exists at the RI in the form of an 18th century marble fireplace set with a Wedgwood plaque, removed at the last minute and presented to the organisation that had its birth in front of it.

But why that particular house? The lasting scientific fame of Sir Joseph Banks rests on the botanical studies that he carried out while voyaging in the South Sea with Captain Cook, but at the time in question he was President of the Royal Society, and thus at the apex of the British scientific establishment. And what of the others present at the meeting? One might think that the founding of a body dedicated to a combination of seeking new scientific knowledge, and then bringing it to the attention of society at large, would have attracted enthusiastic practicing scientists of the day. Not a bit of it.

The attendance list contained a duke, six earls, numerous lords, the Prince-Bishop of Durham, sixteen Members of Parliament, two Directors of the Bank of England and William Wilberforce. Their prime interest in science was the effect that it might have in alleviating poverty. The presence in the same room of this galaxy of high society was due in large part to the efforts of two other men, who were also present: Thomas Bernard, a well-known philanthropist, and an extraordinary North American (the only scientist ever to have been elevated to a Count of the Holy Roman Empire) called Benjamin Thompson, Count Rumford.

Rumford, who we remember as the discoverer of the mechanical equivalent of heat, was an archetypal 'mover and shaker'. He wrote the prospectus, wheedled out the money and drafted the mission statement that appears in the Royal Charter granted in January 1800, and which encapsulates the essence of what the R1 still does: 'to teach by courses of philosophical lectures and experiments, the applications of science to the common purposes of life'.

The next step was to look for premises, and on 5 June of the same year, the first meeting of the Managers was held in the newly acquired house at 21 Albemarle Street, where the RI has been ever since. With the money that flowed in from benefactors (called 'proprietors' in the early years) Rumford directed the building of a lecture theatre on vacant land to the north of Number 21. The present theatre seen on TV every year at the time of the Christmas Lectures (and many other occasions, such as the Reith lectures) occupies exactly the same site as the 1800 one. It also has the same steeply raked semi-circular shape with a gallery above, and sight-lines leading the eye inexorably, not to the lecturer at the podium, but to the bench containing the apparatus for demonstrations.

So, while lecturing about science is at the heart of the RI's ethos, demonstrating science became a tradition from the very beginning. Chemistry arrived early, too, in the person of Thomas Garnett, who lectured on water analysis and minerals, but he was almost immediately eclipsed by a charismatic young man, Humphry Davy, appointed as Director on Rumford's recommendation after the first Director, Thomas Young, resigned. Apart from the Lawrence Berkeley Laboratory under Glenn Seaborg, there can be no other building on the planet that has seen the isolation of so many chemical elements as 21 Albemarle Street under Davy; most of Groups 1 and 2 of the Periodic Table and, at a further remove, chlorine and iodine, were identified there. Davy's lectures on his own discoveries, and on many other topics, brought capacity audiences to the RI lecture theatre, especially the young ladies of Mayfair high society, for whom his Byronic good looks may have held as much allure as his chemistry.

But that was only a beginning: even greater achievements were to come, through the astonishing skill and insight of the young man who left his trade as a bookbinder to come to the RI as Davy's 'chemical assistant': Michael Faraday. Not only was Faraday's entire scientific life passed at the RI and all his discoveries made there, but he, above all, was the person who shaped the organization and its activities into the form that we recognize today. The two series of popular lectures, one for adults and one for children, that he started in 1826 still continue. As a result of increasing commitment from the BBC through the 1990s TV audiences for the young peoples' Christmas Lectures approached 2 million for each of the five lectures. Friday Evening Discourses, conceived by Faraday as 'meetings of an easy and agreeable nature to which members have the privilege of bringing friends, and where all may feel at ease' attract audiences averaging some 300 on 20 Fridays each year, notwithstanding the convention (established later in the nineteenth century) that they are 'black tie' occasions.

At a deeper level than particular series of lectures, the philosophical essence of Faraday's RI remains a potent influence on its ongoing work. That is not a sign of innate traditionalism, but a clear acknowledgement of the validity of his approach. This involves combining, within the same organisation and under the same roof, world class research with a major national outreach programme. And in this way ensuring that the power and excitement of scientific thinking, and its practical results, are brought to wider audiences by the very people most closely involved in shaping them, using live demonstration wherever possible to catch and hold the audience's attention. All these have now become standard features of the science communication business. But, to coin a phrase, you read it here first.

That is not to say, by any means, that nothing much has changed since Faraday's time. Whilst he gave the Christmas Lectures many times from year to year, the really major expansion of lecture-demonstrations for young people was initiated by Sir Lawrence Bragg in the 1950s. Over the past decade, expansion accelerated under the enthusiastic supervision of Richard Catlow and in 1998, it was re-launched with 50 per cent more lectures and increasing numbers now being given outside London. Training for new lecturers and workshops for teachers add further value to the lectures, which bring lines of coaches (though not the same kind as in Davy's time) to Albemarle Street. One of many pleasures in occupying the Director's flat on the second floor is to hear through the window the excited buzz of chatter from the school-children getting out of their coaches, mingled with angry honking taxis trying to get past. The lectures cover all science, with chemistry well represented, and nowadays they take their starting point from the National Curriculum. Nevertheless, they are not pedagogical in the ordinary sense; it's not the RI's job to second guess the schoolteachers. Their flavour is summed up in a remark by George Porter that the RI is not in the educational business, but in the 'inspirational' business. Some 40,000 schoolchildren come to them every year; Faraday would have been amazed.

Research at the RI has always hovered somewhere near the borders of chemistry and physics, although Tyndall's work on ice and atmospheric fine particles had notable consequences for environmental studies. Lawrence Bragg saw the potential (now triumphantly realised) for X-rays to probe biological structures, and George Porter's interest in fast reactions led him to study photosynthesis; the present Director, Susan Greenfield, works on neuro-transmitter molecules. Within the research laboratories in recent years, the focus has been new solids, their synthesis, structure, chemical reactivity and physical properties: reactivity symbolised by catalysis and properties by superconductivity.

Finally, to return to the symbolism of that first meeting on 7 March 1799: science does not belong only to its practitioners, but to the society in which it is embedded and which, nowadays, largely through its taxes, funds it. Throughout its long and glorious history, the RI has sought to combine creating science with communicating it—to young people, professionals, to opinion formers, and to the public at large. Events in recent years have shown how vital such communication is. In the future, it will be even more so. The RI continues to rise to that challenge.

Creating and Communicating Science

If it is a truism that the fabric of modern society is founded on the fruits of science and technology, the consequence must be that it is more important than ever before for the broadest range of the public at large to have some appreciation of how science works, and the kinds of conclusions it reaches. Such understanding has to proceed at two levels: the first is purely professional, in the sense of providing a sufficient number of people with the training needed to operate an advanced technological society. That is the job of the educational system, and is not my theme here. The second level of understanding is more difficult to define and hence to achieve. It is something more pervasive within society: that as many citizens as possible should comprehend the nature of scientific argument and enquiry—what could be called the 'process' of science. That is not so much a matter of spreading knowledge of the scientific principles behind specific issues, such as nuclear power generation or genetic engineering, as of inculcating a feeling (indeed empathy) for the way that new knowledge is uncovered, and hence of the status of scientifically backed statements.

I am delighted to say that I am not alone in these beliefs. In a very welcome development a decade or so ago, the British Prime Minister appointed the first Minister for Science to have a seat in the Cabinet for many years. In advance of announcing his policy White Paper, the Minister William Waldegrave launched a wide consultation exercise, seeking the views of the scientific community, industry and the public at large on what the important issues might be. Among the many points made, it was widely urged on him (by myself among others) that high priority be given to enhancing public awareness of science, engineering and technology, as the makers and arbiters of our lives. For example, I wrote in a phrase that was quoted in the White Paper: 'Any National policy for science and technology must contain, as a necessary foundation, the diffusion among the public at large of an appreciation of what science is'. Such awareness would help the public to know what they could expect of science, and what they could not, and to form soundly argued judgments on matters that require democratically based debate.

One might approach the matter from a narrower point of view: any organisation, be it commercial, industrial or governmental, that spends £1.2 B each year, should (and in most cases does) spend a small fraction of that turnover on explaining what it does and why and how it does it. This should be no less true of the government's research spending. How then is—if I can coin a phrase—the 'public relations of science' organised today? Roughly speaking, it is undertaken in two distinct ways. First, and most straightforwardly, the government agencies responsible for particular fields, such as the Medical Research Council, publicise their activities, and especially their successes, through press releases, brochures, laboratory open days, visiting speaker programmes, etc. Though desirable and valuable, this activity is purely sectorial, and to a certain degree self-justificatory. Therefore, above and beyond this first category of actions, there is a need for programmes that do not suffer from the latter defects, but aim to enhance appreciation of science itself, in a positive spirit but not as a lobby. In Britain three venerable bodies engage in such action: the Royal Society (founded in 1660), the Royal Institution (founded in 1799) and the British Association for the Advancement of Science (founded in 1826). Each goes about its business in different ways, though starting in 1985 they began to act as co-sponsors of a coordinating and facilitating body called COPUS (the Committee on Public Understanding of Science). In the following pages, I want to share with you some of the experience of the Royal Institution in this endeavour, not only because I had the honour to be its Director, but because the way it was set up and the manner in which it carries out its tasks seem to me to carry some valuable lessons.

The United Kingdom is known for its administrative anomalies, and in science the Royal Institution ranks high in that category. Among other things, it houses the oldest continuously operating research laboratory in the United Kingdom, founded in the Age of Enlightenment following the French and American Revolutions. In fact, it was founded by a North American, but a North American who was very much a European, a remarkable man called Benjamin Thompson, otherwise known as Count Rumford. He came by his unusual title as a result of ten years working for the King of Bavaria, reorganising the army. Rumford was a very energetic, inventive man. While in Munich, he devoted himself to useful inventions and, among others, invented a dish which, to this day, can be found in Munich restaurants, called Rumford Soup, which resulted from a research project to discover the cheapest and most nutritious form of sustenance for the poor. He took the matter of the usefulness of science very seriously. That was what he had in mind when, after coming to London, he decided to found a research organisation which would communicate its results to a wider public, a novel concept at that time. It is one which has a very contemporary ring to it—nowadays one would call it a 'research association', that is, the members paid their subscriptions to have the right to learn about the new results and come to the building of the Royal Institution, as it was to be called, to speak with the researchers and attend lectures. So the Royal Institution had a teaching function for the general public in addition to the individual communication of its research results to the subscriber. A most important feature of Rumford's building was what he called the Conversation Room. It still fulfils its original purpose which was, as the name implies, where people can go to talk to each other and where to this day one meets the research students and post-doctoral students over coffee. Rumford's other priority was a lecture theatre, which remains an integral part of the building up to the present time.

In the founding statutes of the Royal Institution, Rumford wrote that its aim was for 'diffusing the knowledge and facilitating the general introduction of useful mechanical inventions and improvements and for teaching by courses of philosophical lectures and experiments the application of science to the common purposes of life'. Apart from a broadening, beyond the word 'mechanical', these phrases encapsulate the essence of what it continues to do till the present day. Before going on to describe how they have been put into practice since 1800, it is worth analysing these words a little more closely. Rumford believed most firmly that a knowledge of science should be deeply embedded in everyday life, and not something separate that was only of intellectual value. For example, his other inventions, based on sound physical principles, included a convector heater and cooking utensils, not to mention a novel cigar lighter. He also believed that those who were creating the new knowledge should be those who communicated it to the public, an obligation which present day scientists should be more widely aware of.

Rumford was never the Director of the Royal Institution (he was much too restless a man for that). He installed a body of Managers and then promptly had a row with them and went off in a huff. Not only in a huff, but with the widow of the eminent French chemist, Lavoisier! Thus, he completed his European tour, having started in Bavaria and passed through London, by ending his life in Paris. In the event, the first Director of the Royal Institution was Thomas Young, who devised the double slit experiment which led him to discover the wave nature of light, and also, in quite a different sphere of intellectual activity, took the first steps to decipher Egyptian hieroglyphs. Young was Director only for a short time when he was succeeded by Humphry Davy, the son of a tin miner, who became famous in London for the quality and interest of his lectures as well as the originality of his research. To this day he remains the person who discovered the largest number of stable chemical elements, in fact most of the alkali metals, the alkaline earth metals and two of the halogens. In addition, he was a charismatic lecturer: people came in large number to the Royal Institution's lecture theatre, and the lectures were even the subject of cartoons in the newspapers (Fig. 1). Even now, the Lecture Theatre of the Royal Institution remains little changed, and I am pleased to say that laughter is still heard there quite frequently.

Not only was Davy the discoverer of a large number of the chemical elements, but he was responsible for one of the most significant inventions in the whole of applied science, the miners' safety lamp. At the Royal Institution we have a beautiful gold cup presented to Davy by the Emperor of Russia in recognition of the number of lives which this invention had saved in the Russian coal mines, truly a potent example of the application

Fig. 1. A public lecture at the Royal Institution: Humphry Davy with the bellows is demonstrating the effect of laughing gas (N_2O). Cartoon by J. Gillray.

of Science to the common purposes of life. Nevertheless, towards the end of his life, Davy was asked what, among all these works, was his greatest discovery: he said 'I have absolutely no doubt that my greatest discovery was Michael Faraday.'

The story of Michael Faraday is among the most romantic in the entire history of science. The son of a blacksmith who lived in a very poor district of south London, Michael left school early and became an apprentice to a bookbinder. The turning point in young Michael's life came the day when one of the customers in the bookshop, a Member of the Royal Institution, gave him a ticket to hear Sir Humphry Davy lecture there on chemistry. Thus it was that he came one evening and sat, as he recorded in his journal, in the centre of the gallery behind the clock. Captivated by the experiments (and by the bangs and smells?), he decided to make his career in science but he did not know how to, because he had no education and he did not know anybody important. He wrote a letter to the President of the Royal Society but, sadly, the President (Sir Joseph Banks) did not reply, so there the matter rested till Faraday had another idea. He wrote a set of notes on Sir Humphry Davy's lectures in beautiful handwriting: we still have this book in the Royal Institution library. He bound it beautifully with his own hands and sent it to Davy as a present, with a letter saying he was so interested by the subject of the lectures that he wished to be employed.

That was the beginning of the story of Michael Faraday as a scientist and of the fifty years that he spent at the Royal Institution. It is probably fair to say that by the sheer range of his discoveries, Faraday was the greatest experimental scientist who ever lived. His stature among Britain's famous may be gauged by the fact that in 1991, to commemorate the bicentennial of his birth, the face of William Shakespeare was removed from the twenty pound bank note and replaced by that of Faraday (Fig. 2). It has been reckoned that, had Nobel prizes existed in the nineteenth century, he should have won six for his discoveries: the laws of electrolysis, the isolation of benzene electromagnetic induction, magneto-optical rotation, diamagnetism and dielectric permittivity. Furthermore, the name Faraday continues to be commemorated by scientists in being applied to

Fig. 2. Faraday on the twenty pound British bank note.

many different phenomena: the unit of electrolysis, the unit of capacitance and, finally, the Faraday effect.

However, it is not on his research discoveries that I wish to concentrate on here. Faraday never forgot the shattering effect on his life that had been brought about by listening to Humphry Davy, and watching the demonstrations that he carried out in front of the astonished audience in the Lecture Theatre of the Royal Institution. As the 'Chemical Assistant', he helped Davy in the preparation of his lecture-demonstrations, and also began to give lectures himself. Becoming more and more convinced how important it was for those who were working in science to spread enthusiasm and deeper knowledge of their work outside the scientific community, in 1826 he began two series of lecture-demonstrations which proved so enduringly successful that both continue up to the present day.

For adult audiences, Faraday conceived the concept of the Friday Evening Discourse. He described the aim and the ambience of these weekly lectures as follows:

> *They are intended as meetings of an easy and agreeable nature to which members have the privilege of bringing friends and where all may feel at ease. It is desirable that all things of interest, large or*

small, be exhibited here either in the library or in the lecture room. The lecture may be long or short, so it contains good matter and, afterwards, everyone may adjourn for tea and talk.

Over the years, almost every scientist of stature has spoken about his work at a Friday Evening Discourse: Rayleigh and Rutherford, the Braggs and Pauling have all been there. And not only scientists; men of letters, poets and philosophers too have been drawn in from time to time. The poet Coleridge, a great friend of Davy, used to attend the Royal Institution in order, as he put it, 'to improve my stock of metaphors', actually quite a good reason why poets might well continue to find interest in them. Of course, since 1826, the format of the Discourses has evolved, though one feature remains constant, the emphasis on lavish illustration through slides, videos and exhibits and above all, where appropriate, demonstrations of the phenomena being expounded. As Faraday said of the scientific profession:

For though to all true philosophers science and nature will have charms innumerable in every dress. Yet I am sorry to say that the generality of mankind cannot accompany us one short hour unless the path is strewed with flowers.

The 'flowers' in question are, of course, the demonstrations and illustrations, a lesson that many of us could profit by today.

In their present day form, the Discourses take place twenty times each year. They are reserved for the Members of the Royal Institution, who pay an annual subscription, and their guests. The sole qualification for becoming a Member is to have an interest in science; although many Members do indeed have some scientific training, many do not, and they are drawn from a wide variety of professions. An additional species of 'flower', to be added to the vivacity of the Discourse itself, is the fact that the evening has very much the character of a *soiree*: dress is formal, a bar is open at the start of the evening in the Council Room, an exhibition on the subject of the Discourse is mounted in the Library and, when the Discourse is over, a buffet is served as part of the price of the ticket. Thus, the occasion is also one at which people can meet one another, and also the lecturer. For example in 1993, we heard, among others, one of the protagonists of cold fusion, Martin Fleischmann, the then newly appointed Director-General of

CERN, Christopher Llewellyn Smith, and the most famous living protagonist of Bach's keyboard music, Rosalyn Tureck.

The Friday Evening Discourses reach a relatively small, though influential, sector of the community. The other programme of lectures established by Michael Faraday in 1826 now reaches a much wider and (some might say) an even more important sector, young people. 'Lectures for a Juvenile Auditory', as Faraday called them, have been given at Christmas time every year since then, except for a brief wartime interruption. Faraday himself gave the Lectures no fewer than seventeen times but, in more recent years, though the Director of the Royal Institution has given them from time to time, it has been the custom to invite others; for example in 1994 we had the 164th annual series, by Professor Frank Close, the Head of the Theoretical Particle Physics Division at the Rutherford Laboratory, on 'The Cosmic Onion'. This was an exploration of matter down to the level of the quarks and leptons, with a view of the Big Bang and the origins of matter. The audience in the Lecture Theatre, with average age about fourteen, is overshadowed nowadays by the enormously larger one accessible through television. Many other celebrated publications have arisen out of the Christmas Lecture series, perhaps the most famous being Faraday's 'Chemical History of a Candle'. The latter, a marvellous piece of scientific exposition, takes as its starting point that humble everyday object to be found on every table in the 1850s, and uses it to uncover most of the principles of chemistry and physics as they were then known: what it is made of, how it burns, how hot the flame is, why it is coloured, and so on. It remains in print to this day, the best-selling edition being in Japanese! To give a flavour of Faraday's beautiful prose style, let me quote the opening of another famous course of lectures he gave 'On the Various Forces of Nature':

> *Let us now consider for a little while how wonderfully we stand upon this world. Here it is we are born, bred, and live, and yet we view these things with an almost entire absence of wonder to ourselves respecting the way in which all this happens. So small, indeed, is our wonder, that we are never taken by surprise; and I do think, that, to a young person of ten, fifteen, or twenty years of age, perhaps the first sight of a cataract or a mountain would occasion him more surprise than he had ever felt concerning the*

> *means of his own existence; how he came here; how he lives, by what means he stands upright; and through what means he moves about from place to place. Hence, we come into this world, we live, and depart from it, without our thoughts being called specifically to consider how all this takes place; and were it not for the exertions of some few inquiring minds, who have looked into these things and ascertained the very beautiful laws and conditions by which we do live and stand upon the earth, we should hardly be aware that there was anything wonderful in it.*

How evocatively he sets the scene for a series of demonstrations of gravity and electro-magnetism; many of the same topics were addressed by Frank Close.

Many other famous scientists have given the Christmas Lectures since Faraday's time. For example, Faraday's successor John Tyndall, perhaps the first natural scientist to devote himself to environmental issues, and the person who first explained satisfactorily why the sky is blue, gave a course on glaciers, and more recently Sir Lawrence Bragg lectured on crystals, while Richard Dawkins, the evolutionary biologist, entitled his lectures 'Growing up in the Universe'. Not only are the lectures reaching a wide audience nowadays through television, but they have been exported beyond the British Isles to South East Asia and, most successfully, to Japan. Figure 3 shows the scene in August 1993 in Tokyo when Professor Charles Stirling lectured on chirality in chemistry and biology under the title 'Left Hand, Right Hand'. The Lectures have also been given from time to time in Singapore and South Korea.

If continuing the tradition of Friday Evening Discourses and Christmas Lectures established by Faraday were the only current contributions the Royal Institution is making to enhancing public awareness of science, it would still be a major endeavour, but might be open to the accusation of remaining static, with one foot in the past. I hope I have said enough to justify the contention that, although established so many years ago, these programmes remain lively and relevant in the present day. However, though maintaining their status as flagships of our enterprise, they have been augmented by many others, and the process of innovation continues. A major development of the 1950s, initiated by Sir Lawrence Bragg,

Fig. 3. Professor Charles Stirling giving the Royal Institution Christmas Lectures in Tokyo, 1993.

was to expand the programme of lecture-demonstrations for young people given in Albemarle Street, so that now they take place several times a week all through the school year. Separate lectures are given for primary and middle schools, and for sixth forms, including Sixth Form Conferences, at which different aspects of a broad subject are treated by three briefer presentations. Recent examples are 'Materials New and Old', with lectures on polymers, superconductors and cement, 'Chaos, Order and Fractals', and 'Energy and the Environment'. Significantly, the fastest growing part of the Schools Lectures Programme is in the primary school age group (8–11), which are regularly oversubscribed. At present, admission to all the lectures is free, although schools have to obtain tickets in advance, so that numbers can be estimated. We are extremely reluctant to introduce even a nominal charge for tickets, as that may turn away children who might benefit most. However, whilst the programme is partly supported by sponsorship from industry and charitable trusts, increasing costs may force us to charge for tickets one day. Information about the lectures is mailed to schools three times a year and, apart from members of the staff of the Royal Institution, many are given by a wide panel of outside lecturers, drawn from universities, industry and schools. More than 30,000 young people each year attend lectures in Albemarle Street, while others have been given outside London.

In parallel with the lecture-demonstrations, Workshops are organised for school teachers in which the content of the lectures is explained in more detail and information given on setting up the demonstrations in a school environment. Finally, it must be emphasised that, whilst the lectures treat subjects that lie within the school curricula, they do not aim to teach: the Royal Institution's function is not to mimic that of the schools. As my distinguished predecessor Lord Porter once said, we are not in the educational business, but the inspirational business. If, as a result of an afternoon spent in the Royal Institution Lecture Theatre, a young person's imagination is captured so that on return to school the curricula comes alive, then our task will have succeeded.

Not only lecturers but also classes given in smaller groups have taken their place in our armoury of activities for young people. Principal among these is the programme of Mathematics Masterclasses, started by popular request after a very successful series of Christmas Lectures by Sir Christopher Zeeman, the first ever given on mathematics. These classes, aimed at able young mathematicians nominated by their schools, have expanded from their beginnings at the Royal Institution to no fewer than twenty-six centres across the country. Another programme beyond the classical lecture format is that of 'Curriculum Enrichment' (RICE) in which, before they arrive at the age when decisions have to be made about examination subject choices or careers, young people are given the opportunity to spend short periods in research laboratories (usually industrial) in their neighbourhoods, to imbibe something of the spirit of the work carried out there.

In these few pages, I have tried to convey how the wealth of activities undertaken by the Royal Institution to raise public consciousness of science, especially among young people, grew out of its history, and in particular the experience of the giants in our past. There can be little doubt that my story is one of success. What lessons, then, can we learn from it? First and most important is to implant a scientific way of thinking in receptive minds, especially those of young people. Second, in pursuing that aim is to recognise that the message comes most potently from those who have been engaged in the scientific adventure themselves, that is, to combine the prosecution of research with exposition to the wider audience. (In this chapter, I have deliberately not expanded on the current research of the Royal Institution's Davy Faraday Research Laboratory: suffice it to say that in recent years the three research groups, totalling some three dozen graduate students, post-doctoral workers and others, published nearly a hundred papers a year, a remarkable rate of productivity). Turning to the means employed, I must emphasise how effective it is to have direct personal contact between the individual who is explaining a topic and the audience—live theatre beats television as a memorable experience. Demonstrations, too, are at the heart of our method. As Sir Lawrence Bragg, himself a master of the lecture-demonstration, said: the difference between being told about a scientific observation and seeing it demonstrated is like learning the character of a foreign country by looking at a map, and by going to visit it.

Finally, let me offer a few quite general thoughts. Not only does the world need to know more about the nature of the scientific endeavour, and its capacity to solve pressing problems, but science will not deserve to flourish unless it can succeed in explaining itself to that large group of people who have never had any professional contact with it. That is true whether one is seeking to capture the imagination of the young, as Davy did for Faraday, or to convince a reluctant Treasury of the support that is needed to continue a line of research. Scientists are members of society, and the fruits of their work underpin and shape it. Society requires and deserves that we enter into dialogue with it: communicating our science is as important as creating it.

The Philosopher's Tree: How Faraday Created Today's Royal Institution

If the Royal Institution could be said to have a patron saint, then that person would have to be Saint Michael: not the familiar symbol of one of the Institution's long-standing Corporate Members (Marks & Spencer) but, of course, Michael Faraday. He it was who, quite apart from all his remarkable discoveries in so many disparate fields of physical science, created the Royal Institution that we still recognize, through the kind of activities it pursues, and the way it goes about them. In a phrase, he set the agenda, of which we are all the inheritors. In the Four Quartets, T. S. Eliot put the relation between past and present in words of quite startling simplicity, as poets will: 'Time present and time past are all perhaps present in time future, and time future contained in time past'.

But the agenda that Faraday established, and pursued so single-mindedly and effectively throughout his life, did not consist only of a programme still faithfully followed by one institution over a 150-year time span. It encompasses a whole approach to the world around us: inanimate, animate, and even social. It contains three elements, and I want to touch on all of them briefly after relinquishing the post which he held for so long and with such unique distinction. The starting point in his approach to the world was vigorous, enthusiastic, imaginative experimentation, asking simple direct questions of nature to discover how the world works: what, in other words, are the 'rules of the game'. The second step was to put the knowledge acquired in front of those who may be most receptive to it, and that means, especially (but not exclusively) young people. The final step is to ensure that society as a whole has these values embedded in it, especially when decisions have to be made on how to proceed with issues where some acquaintance with nature's rules is decisively important (and that, as we know, can mean nearly all issues). We might call that a higher form of education. Faraday gave us striking examples of all three of these elements, and I want to share with you some examples, juxtaposing past and present, and in particular by hearing Faraday's own voice, unfortunately not directly, because sound recording had not been invented in his time, but by what he wrote.

I hope to convince you, too, that among all his other manifold virtues, Faraday had a fine way with words, and it is that which provides me with the title at the head of this text. You may have wondered what a philosopher's tree is. Philosopher was the word commonly used till the middle of the last century to denote what we now call a scientist, but to understand the significance of the word 'tree', consider the following letter written by Michael Faraday at the age of 20, describing how he wished to write:

> *It is my wish, if possible, to become acquainted with a method by which I may write … in a more natural and easy progression. I*

would, if possible, imitate a tree in its progression from roots to a trunk, to branches, twigs and leaves, where every alteration is made with so much ease and yet effect that, though the manner is constantly varied, the effect is precise and determined.

The extracts that follow will enable you to judge how well he succeeded.

As I have indicated, Faraday's programme takes its starting point from carefully, persistently observing the world as it is, probing it, prodding it, and drawing only conclusions that are supported by those observations—that is, by experiment. Faraday described his approach when writing to an old friend, quite late in his life, and his words also serve to remind us of his remarkable beginnings:

I entered the shop of a bookseller and bookbinder at the age of 13, in the year 1804, remained there 8 years and during the chief part of the time bound books. Now it was in these books, in the hours after work, that I found the beginnings of my philosophy. There were two that especially helped me; the Encyclopaedia Britannica, from which I gained my first notions of Electricity and Mrs. Marcet's 'Conversations on Chemistry', which gave me my foundation in that science. I believe I had read about phlogiston etc. in the Encyclopaedia, but her book came as the full light in my mind. Do not suppose that I was a very deep thinker or was marked as a precocious person. I was a very lively, imaginative person, and could believe in the Arabian nights as easily as the Encyclopaedia. But facts were important to me & saved me. I could trust a fact, but always cross examined an assertion. So when I questioned Mrs. Marcet's book by such little experiments as I could find means to perform, and found it true to the facts as I could understand them, I felt that I had got hold of an anchor in chemical knowledge and clung fast to it.

But we should not forget that in the young Michael Faraday's life, looking at the world around him was no chore but on the contrary, was great fun. Imagine, if you will, a rainy evening in London. Two young friends had been spending a weekend afternoon together, and it was time to go home. The following day the 19-year-old Michael wrote to his friend:

Dear Abbott,

Were you to see me, instead of hearing from me, I conceive that one question would be: 'how did you get home on Sunday evening?' I suppose this question because I wish to let you know how much I congratulate myself upon the very pleasant walk (or rather succession of walks, runs and hops) I had home that evening, and the truly Philosophical reflections they gave rise to.

I set off from you at a run and did not stop until I found myself in the midst of a puddle and quandary of thoughts respecting the heat generated in animal bodies by exercise. The puddle, however, gave a turn to the affair and I proceeded from thence deeply immersed in thoughts respecting the resistance of fluids to bodies precipitated into them.

My mind was deeply engaged on this subject, and was proceeding to place itself as fast as possible in the midst of confusion, when it was suddenly called to take care of the body by a very cordial. Affectionate and also effectual salute from a spout. This of course gave a new turn to my ideas and from thence to Blackfriars Bridge it was busily bothered amongst Projectiles and Parabolas. At the Bridge the wind came in my face and directed my attention as well as earnestly as it could go to the inclinations of the Pavement. Inclined Planes were then all the go and a further illustration of this point took place, on the other side of the Bridge, where I happened to proceed in a very smooth, soft, and equable manner for the space of three or four feet. This movement, which is vulgarly called slipping, introduced the subject of friction, and the best method of lessening it, and in this frame of mind I went on with little or no interruption for some time except occasional and actual experiments connected with the subject in hand, or rather in head.

The Velocity and Momentum of falling bodies next struck not only my mind but my head, my ears, my hands, my back and various other parts of my body, and though I had at hand no apparatus by which I could ascertain those points exactly, I knew that it must be considerable by the quickness with which it penetrated my coat and other parts of my dress. This happened in Holborn and from thence 1 went home Sky-gazing and earnestly looking out for every Cirrus, Cumulus, Stratus, Cirro-Cumuli, Cirro-Strata and Nimbus that came from above the Horizon.

The jaunty air of this letter is almost worthy of Gene Kelly in the movie 'Singing in the Rain'. For the young Michael Faraday, experiments could be done anywhere, even at home; in another letter to Benjamin Abbott he describes in triumphant terms his success in making a galvanic cell:

I have lately made a few simple galvanic experiments merely to illustrate to myself the first principle of science. I was going to Knights to obtain some Nickel, and bethought me that they had Malleable Zinc. I enquired and bought some. Have you seen any yet? The first portion I obtained was in the thinnest pieces possible; observe it in a flattened state. I obtained it for the purpose of forming discs, with which and copper to make a little battery. The first I completed contained the immense number of seven pairs of plates and of the immense size of halfpennies each! I, Sir, covered them with seven half-pence and I interposed between seven or rather six pieces of paper, soaked in a solution of Muriate of Soda! But laugh no longer Dear A, rather wonder at the effects this trivial power produced. It was sufficient to produce the decomposition of the Sulphate of Magnesia; an effect which extremely surprised me, for I did not (I could not) have any idea that the agent was component to the purpose.

Such an experiment could easily be done at home to this day by any clever teenager who gets the pieces from a local hardware shop. Nowadays we use electrochemistry at the Royal Institution for rather different purposes, for growing crystals, but still in quite a similar way as can be seen from Fig. 4. The compounds crystallizing from such cells in the basement of 21 Albemarle Street at the present time are superconductors, but made from molecular components. The molecules in question are made in

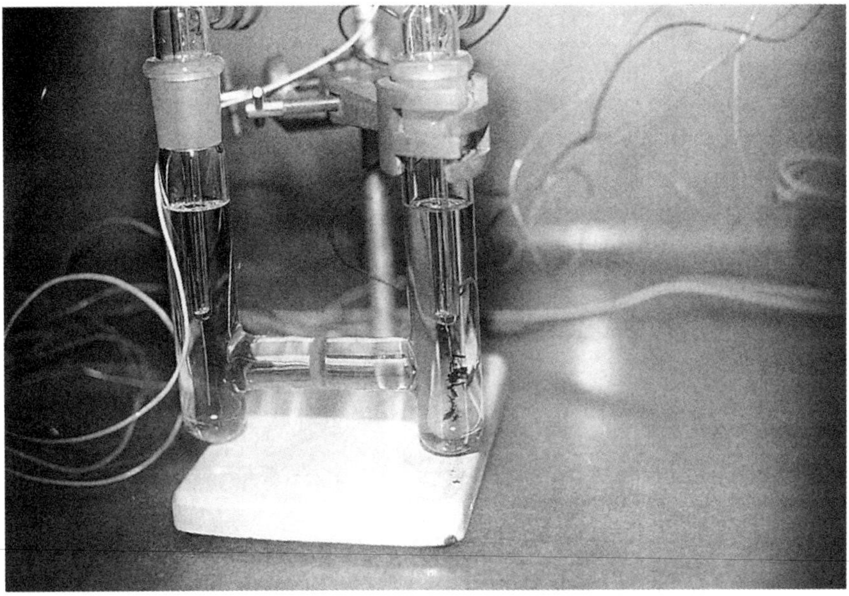

Fig. 4. Electrochemical cells in use at the Royal Institution for growing crystals of molecular charge transfer salts.

a laboratory on the third floor that, while not exactly like the one (Fig. 5(a)) where Faraday made his, maybe not so very different (Fig. 5(b)).

Indeed, the new field of molecular-based magnets and superconductors that is fascinating the members of my own research group at the present time brings to mind another conjunction between past and present in our laboratory. Faraday made much of what he called 'electromagnetic rotations', based on the fact that the current carried by a wire induces a magnetic field in its neighbourhood, as discovered by Oersted. The fact is that the direction of the current determines the polarity of the field, so the arrangement is chiral, i.e. it is a kind of helix. Chirality in nature is a fascinating subject, and indeed a series of Royal Institution Christmas Lecture was devoted to it only a few years ago. However, we have been drawn to look at chirality in crystal structures quite recently, in relation to the molecular superconductors made in our electrochemical cells.

The compounds in question are made up from alternate layers of organic cations, which carry the superconducting current and inorganic anions that

(a)

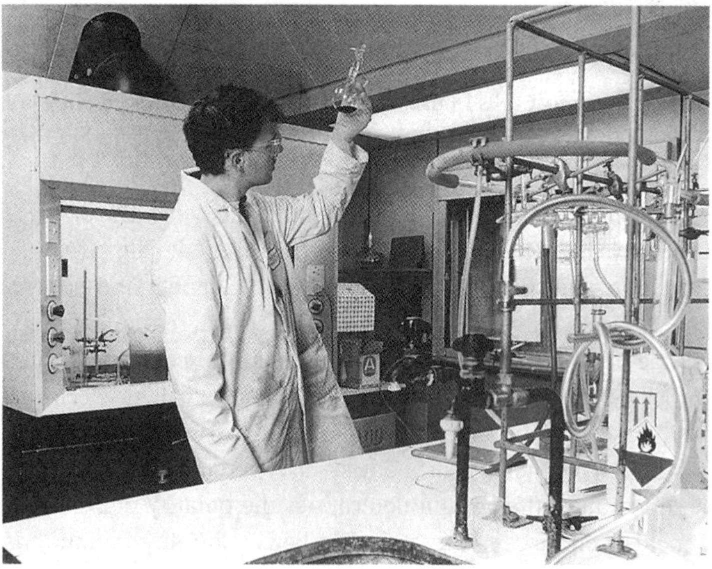

(b)

Fig. 5. Chemistry laboratories in the Royal Institution (a) in Faraday's time and (b) today.

contain unpaired electrons, and hence give rise to paramagnetism. (In passing, it is worth noticing that the words 'cation' and 'anion' themselves were originally coined by Michael Faraday.) In the present context, the important feature of these unusual compounds is that the anions are chiral, actually mimicking three-bladed propellers. What is even more remarkable is that two different materials that differ only in the arrangement of the two kinds of chiral anion have dramatically different properties, one being a superconductor, the other a semiconductor.

It is not very widely known that semiconductivity too was discovered by Faraday, as the following extract from his laboratory notebook makes clear (sulphuret of silver is what we would nowadays call silver sulphide, Ag_2S):

> *21 ST FEBY, 1833*
>
> *Sulphuret of Silver—very extraordinary. At first on piece of glass flask in air, but afterwards in tube, fused into its place in fire. When all was cold conducted a little (by galvanometer) and if quite cold at first conducting power did not increase ... But if battery current strong or if sulphuret continued to increase in conducting power ... the heat rose as the conducting power increased (a curious fact), no other source of heat than the current being present. Yet I do not think it became high enough to fuse the sulphuret ... The whole passed whilst in the solid state. The hot sulphuret seems to conduct as a metal would, and could get sparks with wires at the end and a fine spark with charcoal.*

Figure 6 shows the sketch, from Faraday's notebook, of the simple piece of apparatus for making the measurements.

Perhaps the most perceptive (and, indeed, seminal) of all Faraday's experiments were in the field of magnetism, that enigmatic force, generated spontaneously by only a few substances, but also, as Faraday demonstrated, by electric currents. Figure 7 is his first published picture of the lines of magnetic flux around a bar magnet detected (as school children often do nowadays) with iron filings. Many of Faraday's experiments on this subject were carried out in the 'magnetic laboratory' now reinstated as a museum in the basement of the Royal Institution. In the same small room will be found the

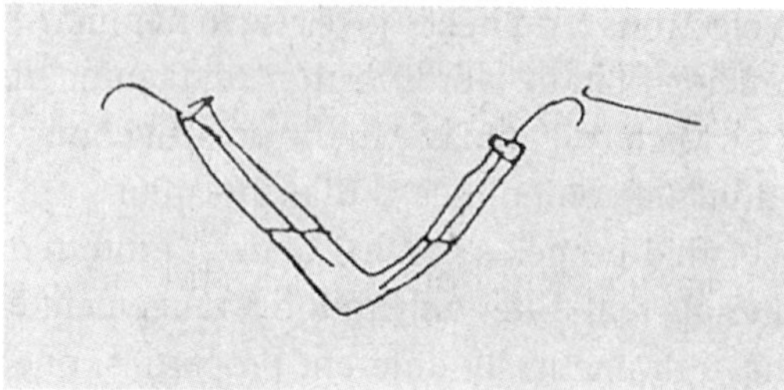

Fig. 6. Faraday's apparatus for measuring the electrical resistance of silver sulphide.

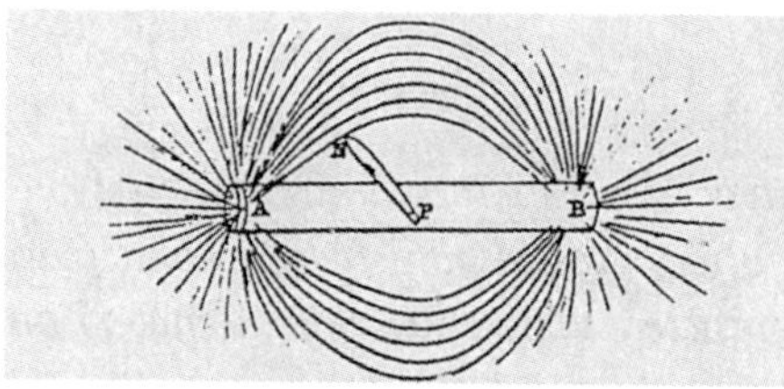

Fig. 7. The first published picture (1839) of the lines of magnetic flux around a magnet.

'great electromagnet' that he used. Nowadays, we use something a bit more elaborate to investigate the magnetic properties of the compounds that we make, as may be seen from Fig. 8, which shows the sensitive magnetometer based on a superconducting magnet, and a detector known as a SQUID (not seafood, but a 'superconducting quantum interference device'). Why should we still be studying magnetism? Surely we know all about magnets, and what does a chemist have to do with magnets anyway?

The magnet in Fig. 7 was a bar of iron, but nowadays chemists can make magnets that are transparent, such as the one shown in Fig. 10, which was discovered in my research group some 20 years ago, or even soluble, by building the solid out of molecules. Such efforts have also led to magnets that do not contain any metal atoms at all, only the elements C, H, N, such as the one whose molecular structure is shown in Fig. 9, although up till now they have only been found to be magnetic at very low temperatures.

Fig. 8. Magnetometer using a superconducting magnet.

Fig. 9. An optically transparent ferromagnetic material.

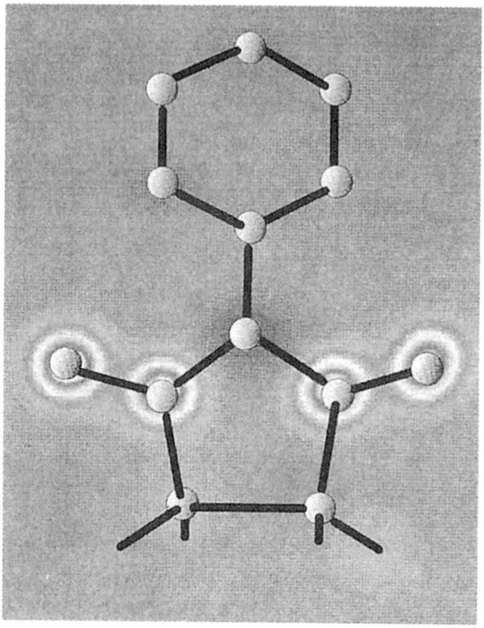

Fig. 10. Molecular structure of an organic ferromagnet.

One reason for being interested in transparent magnets goes back to one of Faraday's most seminal experiments (as always, a very simple one), which he described in his notebook:

> *A piece of heavy glass. . .being silico-borate of lead, and polished on the two shortest edges, was experimented with. It gave no effects when the same magnetic poles or the contrary poles were on opposite sides (as respects the course of the polarized ray)—nor when the same poles were on the same side, either with the constant or intermitting current—BUT, when contrary magnetic poles were on the same side, there was an effect produced on the polarized ray, and thus magnetic force and light were proved to have relation to each other. This fact will most likely prove exceedingly fertile and of great value in the investigation of both conditions of natural force.*

It is worth drawing attention to the word BUT, written in capital letters and underlined several times, in contrast to the rest of the text, written in his usual immaculate copperplate handwriting. Clearly, he realized at once how significant the result was, as we can also observe from his final comment, one of the most remarkable pieces of understatement in the history of science.

Only a very small minority of solids are spontaneously magnetic, but another great discovery of Faraday's was that all substances have the property of what he called diamagnetism, i.e. that the lines of magnetic flux diverge instead of converging, within the material. His experiments designed to illustrate the universality of the property lead to some amusing asides in his notebooks:

> *It is curious to see such a list as this of bodies presenting on a sudden this remarkable property, and it is strange to find a piece of wood, or beef, or apple, obedient to or repelled by a magnet. If a man could be suspended, and placed with sufficient delicacy in the magnetic field, he would point equatorially.*

Many years after Faraday carried out his experiments on diamagnetism, it turned out that this property is one of those characterizing the superconducting state, about which we spoke earlier. Indeed, a superconductor is a perfect diamagnet, which brings about the possibility of levitating objects by posing an array of magnets above a superconducting plate. At least up till now, superconductivity remains inherently a low temperature property, but with the advent of materials showing the property above the boiling point of liquid nitrogen, some quite spectacular demonstrations become possible (Fig. 11). Indeed, several years ago, when giving a Friday Evening Discourse on Superconductivity, I was able (quite appropriately) to levitate a bust of Michael Faraday! Magnets made from superconducting wire are now very big business as high magnetic fields now form the basis of many advanced measurement and diagnostic techniques. The one that has made the greatest impact on the public is body imaging, called magnetic resonance imaging, which is now found in many hospitals. The quality of the images obtained can be gauged from Fig. 12, which holds a special interest for the author, whose vertebrae they are!

Fig. 11. Magnet levitating above a superconducting ceramic in liquid nitrogen.

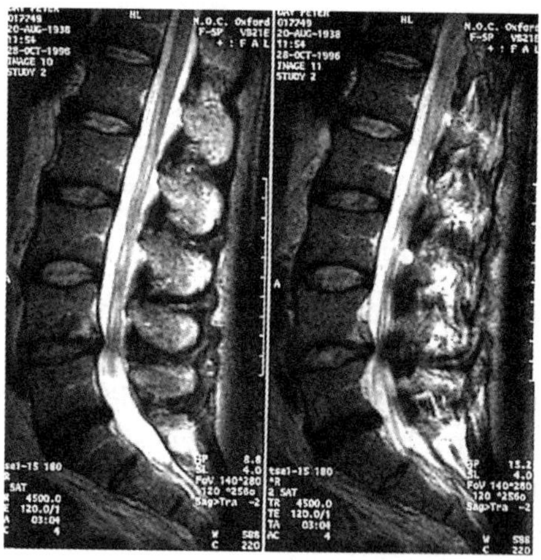

Fig. 12. The author's backbone seen by magnetic resonance imaging.

Returning to the theme of Faraday's life and work, we may ask what way of life it was that enabled him to accomplish all these great discoveries. The answer is, a very constrained one (constrained, that is, in a geographical and social sense, not of course in an intellectual one). The strength of his attachment to the Royal Institution is nicely brought out in one of his letters to the American physicist Joseph Henry:

> *Your accounts of your transits over the world, and changes in the position of your family, almost startle and shame me, for I feel as*

if I could have shown none of the energy and perseverance which carries you through all these things. I have been here so attached to the Royal Institution that I feel as if I were a limpet on a rock, and that any chance which might knock me from my position would leave me but little hopes of attaching myself anywhere again.

Having have seen his laboratory, it is pertinent to look at his office. Fortunately for us, a friend of the Faradays, Harriet Moore, was a watercolour painter (her work still hangs upstairs in the Director's Flat), and Fig. 13 juxtaposes the 1850s with 1998. Eagle eyes may notice that some of the furniture remains the same, although the quantity of books and papers has grown.

A key to Faraday's success in accomplishing so much in this modest domestic environment was his preoccupation with time, and how to use it to best effect. This trait developed earlier in life, as we see from the following letter to his teenage friend Benjamin Abbott:

Dear Abbott,

What is the longest and the shortest thing in the world: the swiftest and the most slow: the most divisible and the most extended: the least valued and the most regretted? Without which nothing can be done; which devours all that is small and gives life and spirits to everything that is great? It is that, good Sir, the want of which has till now delayed my answer to your welcome letter. It is what the Creator has thought of such value as never to bestow on us mortals two of the minutest portions of it all at once. It is that which, with me, is at the instant very pleasingly employed. It is Time.

Later in life he was equally trenchant about time wasting on official business, especially committees (we may well sympathize!):

With respect to committees, as you would perceive I am very jealous of their formation. I mean working committees. I think business is always better done by few than by many. I think also the working few ought not to be embarrassed by the idle many and, farther, I think the idle many ought not to be honoured by association with the working few. I do not think that my patience has ever come nearer

(a)

(b)

Fig. 13. The Director's study at the Royal Institution (a) in Faraday's time and (b) in 1998.

to an end than when compelled to hear (in the examination of witnesses etc. in committee) the long rambling malapropros enquiries of members who still have nothing in consequence to propose that shall advance the business. But in all this too, I will promise to behave as well as I can.

By a natural progression, work in the laboratory also led to the lecture theatre, not so much the formal professional lecture as the popular exposition. For Faraday, as for his successors at the Royal Institution, telling people about science is as important as doing it in the first place. And the means that he used were the same as the tools of his research: direct simple experiments and clear, approachable explanations. Nowhere is that more evident than in the Christmas Lectures, now as much as then. 'Then' is symbolized by the famous picture of the 1862 lectures, given before the Prince Consort and the two royal princes (Fig. 14(a)). 'Now' is symbolized by the ubiquitous presence of television (Fig. 14(b)), although the two occasions are strikingly similar.

One of the most eloquent of his hymns in praise of science comes, not from his most famous series of Christmas Lectures, 'The Chemical History of a Candle', but the less well known 'On the Various Forces of Nature', which concentrates on physics. It has already been quoted on p. 14, but in the following, one can almost hear the voice as he introduces himself:

I shall here claim, as I always have done on these occasions, the right of addressing myself to the younger members of the audience. And for this purpose therefore, unfitted as it may seem for an elderly infirm man to do so, I will return to second childhood and become, as it were, young again amongst the young.

Two extremely simple demonstrations about cohesive forces will serve to illustrate his approach. The first uses a towel as a syphon:

When you wash your hands you take a towel to wipe off the water; and it is by that kind of wetting, or that kind of attraction, which makes the towel become wet with water, that the wick is made wet with the tallow. I have known some careless boys and girls (indeed, I have known it happen to careful people as well) who, having washed

(a)

(b)

Fig. 14. Royal Institution Christmas Lectures (a) in 1862 and (b) in 1993.

their hands and wiped them with a towel, have thrown the towel over the side of the basin, and before long it has drawn all the water out of the basin and conveyed it to the floor, because it happened to be thrown over the side in such a away as to serve the purpose of a syphon.

The second demonstrates how adding salt to water lowers its freezing point:

I remember once, when I was a boy, hearing of a trick in the country alehouse. The point was, how to melt ice in a quart-pot by the fire and freeze it to the stool. Well, the way they did it was this: they put some pounded ice in a pewter pot and added some salt to it, and the consequence was that when the salt was mixed with it, the ice in the pot melted (they did not tell me anything about the salt), and they set the pot by the fire just to make the result more mysterious. And in a short time the pot and the stool were frozen together, as we shall very shortly find it to be the case here. And all because salt has the power of lessening the attraction between the particles of ice. Here you see the tin dish is frozen to the board. 1 can even lift this little stool up by it.

In the more famous 'Candle' series we also have instances of the clearest kind of exposition, not all of whose lessons have been learnt up to the present day. For example, take the eloquent account of respiration by plants:

All the plants growing upon the surface of the earth, like that which I have brought here to serve as an illustration, absorb carbon. These leaves are taking up their carbon from the atmosphere, to which they have given it in the form of carbonic acid, and they are growing and prospering. Give them a pure air like ours, and they could not live in it; give them carbon with other matters and they live and rejoice. This piece of wood gets all its carbon, as the trees and plants get theirs, from the atmosphere, which, as we have seen, carries away what is bad for us and at the same time good for them: what is disease to the one being health to the other. So are we made dependent, not merely upon our fellow-creatures, but upon our fellow-existers, all

Nature being tied together by the laws that make one part conduce to the good of another.

As a recent television programme revealed by interviewing newly graduated students from one of the world's most prestigious academic institutions, Massachusetts Institute of Technology, this simple lesson is still not universally acknowledged, even among the most highly educated. Such misconceptions lead us to wonder very seriously about the state of education of the public at large. In his time, too, Faraday had cause to wonder at the state of education of the public at large about the laws of nature, especially in the 1850s when a craze of spiritualism, exemplified by table turning swept London. He even wrote a letter to The Times about it:

Permit me to say that I have been greatly startled by the revelation which this purely physical subject has made of the condition of the public mind. No doubt there are many persons who have formed a right judgment or used a cautious reserve. But their number is almost as nothing to the great body who have believed and borne testimony, as I think, in the cause of error. I do not here refer to the distinction of those who agree with me and those who differ. By the great body, I mean such as reject all consideration of the equality of cause and effect, who refer the results to electricity and magnetism, yet know nothing of the laws of these forces, or to attraction, yet show no phenomena of pure attractive power—or to the rotation of the earth, as if the earth revolved around the leg of a table—or to some unrecognised physical force, without inquiring whether the known forces are not sufficient—or who even refer them to diabolical or supernatural agency, rather than suspend their judgment, or acknowledge to themselves that they are not learned enough in these matters to decide on the nature of the action. I think the system of education that could leave the mental condition of the public body in the state in which this subject found it, must have been greatly deficient in some very important principle.

This concern for the state of public education about what Faraday calls 'cause and effect' is still with us, but it led him to examine what role a knowledge of science can bring—appropriately enough in a Friday Evening

Discourse on the relation between basic science and the electric telegraph. (Shades of today's debates about pure and applied science!) His words are more eloquent than I could muster, so I will quote again:

> *If the term 'education' may be understood in so large a sense as to include all that belongs to the improvement of the mind, either by the acquisition of the knowledge of others or by the increase of it through its own exertions, we learn by them what is the kind of education science offers to man. It teaches us to be neglectful of nothing, not to despise the small beginnings, for they precede of necessity all great things in the knowledge of science, either pure or applied. It teaches a continual comparison of the small and great, and that under differences almost approaching the infinite: for the small as often contains the great in principle as the great does the small, and thus the mind becomes comprehensive. It teaches to deduce principles carefully, to hold them firmly, or to suspend the judgment: to discover and obey law, and by it to be bold in applying to the greatest what we know of the smallest. It teaches us first try tutors and books to learn that which is already known to others, and then by the light and methods which belong to science to learn for ourselves and for others, so making a fruitful return to man in the future for that which we have obtained from the men of the past.*
>
> *The beauty of electricity, or of any other force, is not that the power is mysterious and unexpected, touching every sense at unawares in turn, but that it is under law, and that the taught intellect can even now govern it largely. The human mind is placed above, not beneath it, and it is in such a point of view that the mental education afforded by science is rendered super-eminent in dignity, in practical application, and utility.*

That people who believe themselves educated should know little about science was a source of disappointment and disquiet to Faraday, as we see from the evidence he gave to the Royal Commission on Education in 1862:

> *The phrase 'training of the mind' has to me a very indefinite meaning. I would like a profound scholar to indicate to me what he*

> *understands by the training of the mind: in a literal sense, including mathematics. What is their effect on the mind? What is the kind of result that is called the training of the mind? Or what does the mind learn by that training? It learns things, I have no doubt. But does it learn that training of the mind that enables a man to give a reason in natural things for an effect which happens from certain causes? Or why in any emergency or event he does (or should do) this, that, or the other? It does not suggest the least thing in these matters. It is the highly educated man that we find coming to us again and again, and asking the most simple question in chemistry or mechanics: and when we speak of such things as the conservation of force, the permanency of matter, and the unchangeability of the laws of nature, they are far from comprehending them, though they have relation to us in every action of our lives. Many of these instructed persons are as far from having the power of judging these things as if their minds had never been trained.*

Or, in more colloquial vein, in answer to one of the Commissioners, you can hear the irony in his voice:

> *I am not an educated man, according to the usual phraseology and therefore can make no comparison between languages and natural knowledge, except as regards the utility of language in conveying thoughts. But that the natural knowledge which has been given to the world in such abundance during the last 50 years should remain untouched, and that no sufficient attempt should be made to convey it to the young mind growing up and obtaining its first views of these things, is to me a matter so strange that I find it difficult to understand. Though I think I see the opposition breaking away, it is yet a very hard one to overcome. That it ought to be overcome, I have not the least doubt in the world.*

This kind of education, acknowledging the relation between cause and effect, is borne in on all of us from our earliest years as children, and becomes the basis of an intuitive feeling for how the world works. What science does is to inform us, after diligent enquiring and sifting of evidence, that such intuitions are not always valid. As intuition is a poor guide, we come back to systematic science, to observing the world, prodding it and summing up what we find in general laws, not immutable but provisional, although valid till altered:

> *The laws of nature, as we understand them, are the foundation of our knowledge in natural things. So much as we know of them has been developed by the successive energies of the highest intellects, exerted through many ages. After a most rigid and scrutinizing examination upon principle and trial, a definite expression has been given to them. They have become, as it were, our belief or trust. From day to day we still examine and test our expressions of them. We have no interest in their retention if erroneous. On the contrary, the greatest discovery a man could make would be to prove that one of these accepted laws was erroneous, and his greatest honour would be the discovery.*
>
> *Let us go out into the field and look at the heavens with their solar, starry, and planetary glories; the sky with its clouds; the waters descending from above or wandering at our feet; the animals, the trees, the plants; and consider the permanency of their actions and conditions under the government of these laws. The most delicate flower, the tenderest insect, continues in its species through countless years, always varying yet ever the same. These frail things are never-ceasing, never changing, evidence of the law's immutability.*

What a man, what a patron saint, what an inspiration!

A Special Friday Night

When I was growing up as a boy in a small village in Kent in the 1940s, Friday night was bath night. At the Royal Institution, since 1826, we have known differently: Friday night is Discourse night. In fact, the database of Discourse speakers and titles, maintained now on computer, informs me that since the first Friday Evening Discourse given by Michael Faraday on 'Catchouc' on 3 February 1826, there had been no fewer than 3,349 up to the one given on 18 March 1994. Even among such a large number, that evening's Discourse was an especially notable one, for two reasons. First, it marked the beginning of the National Week of Science, Engineering and

Technology, an event instigated by the Office of Science and Technology as part of the campaign to enhance public appreciation of these matters, foreshadowed in the previous year's Government White Paper 'Realising Our Potential'. Second, to the best of my belief, it was the first time in the 168-year history of the Friday Evening Discourses that a serving member of the Cabinet, and Minister responsible for Science, had given one. For both of those reasons, it was an honour and a pleasure to welcome the Right Honourable William Waldegrave MP, Chancellor of the Duchy of Lancaster and Minister of Public Service and Science, to the Royal Institution.

The Friday Evening Discourses were arguably the first sustained endeavour anywhere in the world to help the public towards a better understanding of the scientific approach to our world, although the Royal Institution's efforts in this direction go back even further, to Humphry Davy's popular lectures on chemistry. One of the lesser known facts of London life is that the first ever street to be officially designated 'one way' is the one in which the Royal Institution is situated. The reason why it was so designated is perhaps even less well known: it is a tribute to the huge popular success enjoyed by Davy's lectures. Mayfair high society came in such numbers in their carriages that the street was blocked, and a traffic control system had to be installed. So much a part of the social scene were they that cartoons of them appeared in the press (Fig. 1, p. 10). What a challenge it would offer to achieve comparable press coverage for a scientific lecture today.

At the beginning of the nineteenth century, science was considered an integral part of the nation's cultural life and, for example, Humphry Davy was a close friend of both Wordsworth and Coleridge, both of whom stood at the rostrum in the Royal Institution's Lecture Theatre. Indeed, in a flight of poetic fancy uniting art and science, Coleridge wrote:

> *'Water and flame, the diamond, the charcoal... are convoked and fraternised by the theory of the chemist... It is the sense of a principle of connection given by the mind, and sanctioned by the correspondency of nature... If in a Shakespeare we find nature idealised into poetry, through the creative power of a profound yet observant meditation, so through the meditative observation of a Davy... we find poetry, as it were, substantiated and realised in nature: yea, nature itself disclosed to us... as at once the poet and the poem!'*

Would it were so today. Indeed, during the consultation that led up to the 1993 White Paper, many of us urged that improved public appreciation of the role of science, engineering and technology in our society was an essential underpinning of any national strategy in these matters. So we were delighted at the emphasis given to these concerns in the White Paper itself—and the Royal Institution, in particular, was especially gratified to see its memorandum directly quoted with approval. So, at the outset of what has to be a sustained and continuous programme of endeavour—to bring science back to the centre of the stage—why, I always think, to quote the founder of the Salvation Army, should the Devil have all the best tunes?

To conclude, the presentations of science at the Royal Institution have always included demonstrations, and it would not have done justice to that tradition, on the occasion of the Minister's visit and the first National Science Week, not to introduce at least a couple which, together, make a point central to the concerns of the Science Week. Both were in the nature of 'executive toys', small objects to beguile the eye. One dated from 1821, the other from 1987. The earlier was, in essence, the world's first electric motor, made in the house of the Royal Institution. Think what has sprung from it! The more recent was the first high temperature superconductor. Its consequences we do not yet know. But imagine flywheels, bearings or other electrical and electronic devices. In a few years time perhaps, as Faraday famously replied to an enquiry about the usefulness of one of his discoveries, 'government will tax it!' And why not? Because that will be a sign that it has left the realm of basic science and entered that of industrial exploitation. If it is to attract a tax, it will be as a result of becoming the basis of some device that has fulfilled a need, and created a profit. The world will have been improved, the exchequer strengthened, and the public will have become aware of one more scientific marvel.

Christmas Lectures in Japan

Arguably, Michael Faraday was the most creative experimental scientist who ever lived: the first electric motor, transformer and dynamo, the discovery of benzene, and the connection between light and magnetism all testify to his brilliance in the laboratory. But he was more: a charismatic lecturer who captivated audiences with simple demonstrations of chemical and physical principles and a far-sighted thinker about the role of science in society.

In fact, Faraday was the first person in history to take seriously what we now call (perhaps a bit pompously) 'the public understanding of science' as an issue of national importance and in his hands the Royal Institution was the first organisation to devote itself to this task. In fact, one of Faraday's innovations not only maintains its hold on popular esteem in the United Kingdom right up to the present day, but has turned into a powerful vehicle for bringing the British way of doing science to a wider public abroad, especially in Japan.

Pursuing his notion of bringing science to a wider public, in 1826 Faraday started two series of lectures at the Royal Institution: both continue to this day. 'Friday Evening Discourses' were to be aimed at an adult audience, while the 'Christmas Lectures' were for young people (or what he called 'a juvenile auditory'). Since that time there have been no fewer than 184 such annual series of lectures for young people. Surely there can be few institutions in British public life that have proved so consistently popular? Faraday himself gave the lectures no fewer than seventeen times, but nowadays the custom is to invite a different lecturer each year. One of the original features of the lectures continues though, and certainly contributes to their appeal: the lavish use of demonstrations to illustrate the points made.

Although they are still given in the same Lecture Theatre at the Royal Institution where Faraday himself lectured, in recent years the audience for the Christmas Lectures has grown enormously in two respects. First is through the medium of television. From 1966 to 2000, the lectures were broadcast by BBC2 and since then by Channel 4, so each one is heard by about one million people—not just teenagers of course, but the proverbial 'eight to eighty'. In fact, the viewing figures are among the highest for television science programmes, and the Royal Institution lectures have become as much a part of the Christmas broadcasting scene as the Festival of Nine Lessons and Carols from King's College, Cambridge. From 1988 to 1997, they were produced with sponsorship from Shell and since then from GSK. However, the second way in which the audience for the lectures has expanded is, perhaps, even more exciting.

During the Directorship of Lord Porter, the Christmas Lectures were given from time to time in South East Asia, but in 1987, for the first time,

Professors John (now Sir John Meurig) Thomas and David Phillips took them to Japan, where they were instantly successful. Japan has always looked with fascination at the scientific history and culture of Britain: Faraday's book 'The Chemical History of a Candle', based on the Christmas Lectures he gave in 1848, is a bestseller among Japanese schoolchildren. The live repeats of the lectures given in London were a sell-out. Since then, they have been given every year at two locations: one outside Tokyo in halls much bigger than the one in Albemarle Street to audiences of over a thousand young people at a time (Fig. 15). Simultaneous translation is used, and the equipment made specially in our own workshops is shipped to Japan and reconstructed.

Officials in the Tokyo office of the British Council have proved invaluable intermediaries and facilitators, not least in establishing sponsorship by the leading Tokyo daily newspaper Yomiuri Shimbun, said to have the largest circulation of any in the world. Relations with Yomiuri have proved excellent, and on visits to Japan I and my successor as Director, Susan Greenfield, were able to reach agreement with them for sponsorship for some years to come.

Fig. 15. Professor Charles Stirling giving the Christmas Lecture in Japan.

In summer 1994, to mark the fifth year of the Yomiuri sponsorship, there were extra events, including popular lectures about the Royal Institution and discussion groups about the public understanding of science, while in 1999, to mark the RI's bicentenary, more such events were organised. So a showcase of British scientific history, and a window on contemporary British science, have been opened. As Anglo-Japanese relations burgeon and deepen, the Royal Institution is doing its part by bringing Faraday's legacy to a new audience of young people on the other side of the globe.

chapter 2

Conversation Rooms

If you walk northwards away from Piccadilly, in London's West End, along Albemarle Street you will soon find yourself passing an imposing building fronted with large stuccoed columns. This is the Royal Institution. But 'Royal Institution of what?' you may ask (as taxi-drivers often do); actually 'The Royal Institution of Great Britain', to give it its full name, although that gnomic appellation reveals very little. In fact, it is one of the most illustrious addresses in British science and the oldest continuously operating research laboratory in the world, having been founded by the efforts of Benjamin Thompson, Count Rumford, in 1799. My reason for drawing attention to it here, though, is not so much to point to the truly astonishing number of discoveries made behind those august columns as to use it as a metaphor for introducing some thoughts about how progress in understanding the natural world actually takes place in practice. If a synonym for the natural sciences is 'validated knowledge', then the necessary prerequisite for validating is sharing. That is followed by debate and sometimes by disagreement but ultimately by consensus, albeit of a provisional kind, pending arrival on the scene of more facts or neater explanations. Of some interest, therefore, are the kinds of venue where the sharing and debate take place. In a real sense, they are the crucibles out of which new science is forged. I want to mention just a few that have seemed to me especially effective from my own experience.

Fig. 16. The Conversation Room at the Royal Institution in 1809 as seen by Thomas Rowlandson.

Looking through the tall sash windows on the ground floor of the Albemarle Street front of the Royal Institution, you will see, not a laboratory with benches and apparatus (those are hidden away upstairs and in the basement) but a sober elegant room lined from floor to ceiling with books. Clearly it must be some kind of library. Well, not exactly, because a closer look will reveal groups of chairs around tables mostly occupied by young people. But they are not so much reading the books as drinking coffee and talking animatedly. In a library, the most prominent sign is usually the one that says 'silence'. Here, it is the reverse for this is called the 'Conversation Room'.

When Rumford established the Royal Institution, there was precious little government or corporate involvement in scientific research so he approached a number of wealthy individuals who, in return for donating money to purchase the premises at 21 Albemarle Street and pay the wages of the staff, became members of what was essentially a private club. Indeed, for the first 10 years or so, till Humphry Davy revised the statutes, they were even called 'proprietors', for that is exactly what they were. The Conversation Room was an idea of Rumford's to provide a space where people interested in science (and why would the proprietors have put up the money if they were not?) to meet one another to discuss the latest developments and also talk to the researchers themselves. Which (give or take the changes in habits and custom over the last 200 years) is pretty much what it is still used for today.

Since the present day Members of the Royal Institution are mostly busy people pursuing their own careers, most of the figures you will see through the ground floor windows of 21 Albemarle Street are PhD students or post-doctoral researchers, with a few seniors who are Professors leading the research teams. But my reason for drawing attention to the goings-on in the Conversation Room is not to emphasise how long-lived some traditions still are, even less to draw attention to the perennial garrulousness of the young, but to correlate all this chatter to the fact that, over 200 years, 21 Albemarle Street has been one of the places most productive of scientific novelty on the entire face of the planet. 'Correlated' is a provocative word, so let me add some more examples to support my proposition that conversation and productivity in science do go hand in hand.

Some fifty miles northwest of the Royal Institution's Conversation Room lies another such room in a famous centre for science which, if not quite so venerable as Rumford's salon, at least dates back architecturally to the middle of the nineteenth century. The Abbot's Kitchen in Oxford was originally designed as the chemical laboratory of the University. As such it formed part of a uniquely imaginative and ambitious plan for a centre grouping all the natural sciences within a single edifice called simply the University Museum, built (in what was then splendid isolation) in a corner of the University Parks. But why should it be called the 'Abbot's Kitchen'? To explain that, we have to look more closely at the arrangement of the building and the ethos of the 1860s when it was constructed.

The conception of the Museum was sweeping; to give each Professor an office and laboratory, with a large and a small lecture theatre shared by all. Up to the iconoclastic 1960s both lecture theatres retained their original semi-circular form, modelled in fact on the amphitheatre that was Rumford's first building project at the Royal Institution 60 years earlier. All these facilities were grouped on two levels around a large square courtyard, glazed over in the manner of the Crystal Palace. The latter was intended as a museum, in the sense that it served as repository for collections of specimens (flora, fauna, minerals and so on) to provide the basis for practical instruction. (It has to be recalled that all this took place in the heyday of taxonomy as a bedrock of botany, entomology and so on.) At the outset, only one branch of practical science did not have its place within the hollow square of the main building, and that was chemistry. Why? Most probably because it carried the whiff of danger through fire and explosion—or simply the whiff *tout court*.

Having decided that chemistry should be banished to a free-standing edifice quite separate from the rest of the natural sciences, the question arose as to what an appropriate architectural model might be. Bear in mind that the remainder of the building had come under close scrutiny by Ruskin, for whom the 'middle gothic' or 'pointed' style represented the apotheosis of human achievement in stone, and that even the capitals and columns in the Museum courtyard were exhibits, the former carved into representations of British flora and the latter each composed of a different British marble. But what about chemistry?

Apart from Roger Bacon in his monastic cell at Oxford castle, there was little medieval precedent for the kind of activities that might go on in a nineteenth century chemistry laboratory; but how about cookery? Chemists have always resented suggestions that their delicately orchestrated activities should bear such a comparison but in 1860 it may not have been so wide of the mark. So, a kitchen? With communal meals a central feature of medieval monastic life, kitchens can be found in many ruined abbies. An especially well-preserved specimen exists at Glastonbury and that is what the architect of the Oxford Chemistry Laboratory chose.

The Abbot's Kitchen at Glastonbury is octagonal, with four large fireplaces occupying alternate sides and a pyramidal roof, crowned at the apex with a turret. The latter disguises a louvred vent designed to let out the smoke and cooking smells—altogether an admirable model for a nineteenth century chemical laboratory. In the Oxford realisation of this precedent, the fireplaces equipped for roasting were transformed into glass-fronted fume-cupboards, though fire was still called into service in the form of gas jets along the base of the openings designed, not for heat or light as such, but to create a convection current in the chimney. Work-benches radiated from the centre while over all arched six curved black painted beams, converging at the top towards a cupola that could be opened to exhaust the chemical fumes.

With passing time, and the ever-increasing needs for teaching and research accommodation (not to mention an appetite for building by

Fig. 17. The Abbot's Kitchens. Top: Glastonbury; bottom: Oxford.

successive generations of science professors verging on the megalomaniac), the originally free-standing Abbot's Kitchen was gradually surrounded by a motley assortment of other edifices in a variety of architectural styles ranging from high Victorian Pugin-esque gothic to what Osbert Lancaster called 'Pont Street Dutch'. Though the competition is strong, perhaps the most banal and bombastic is a dreary cliff-like building on South Parks Road. It is constructed like a castle from rusticated stone, set around a small dank airless quadrangle entered from the road through a barrel-vaulted archway just too narrow to allow easy access by cars. As well as boasting the largest number of corridors per unit of useable floor-space in any Oxford building, it had the dubious distinction of having been designed in 1938, when the New Bodleian Library was considered the height of architectural chic but, because of the Second World War, not erected till ten years later. Thus it contrived to combine technical arrangements that were already out of date when it was built with one of the most inconvenient ground plans ever conceived for a laboratory. Over the archway, where the dressed Portland stone might have invited some improving inscription incised into it (probably of classical origin like the one on Rhodes house opposite), bronze letters announced the less-than-memorable message INORGANIC CHEMISTRY.

So, by the end of the 1950s the magnificent architectural celebration of chemistry represented by the Abbot's Kitchen had fallen on hard times. Not only was this the result of the stylistically mongrel collection of buildings that surrounded it, but also the use that it was then put to. In short, it had become a lumber-room. Nevertheless, the 1960s were a period of expansion, optimism and new departures. In 1963, for the first time in its history, the University appointed a Professor of Inorganic Chemistry. Up till then the three branches into which chemistry was traditionally divided were led by just two Professors, whose titles (Waynfleet and Dr Lee's) announced their venerable standing. Thus one of three branches was placed under the day-to-day supervision of a mere Reader and, in the 1960s, that was Inorganic Chemistry. Ultimate authority rested with one of the so-called 'statutory' chairs, at that time the Dr Lee's Professor. Its holder was the eminent physical chemist, Nobel Prize winner (and—in those far off days of broad scholarship—former President of the Classical Association), Sir Cyril Hinshelwood.

It was widely believed (perhaps even correctly) that Hinshelwood held the view that inorganic chemistry was not a real subject, in consequence of which he used posts allocated to it to find appointments for a number of able young physical chemists for whom no vacancies could be found in the Physical Chemistry Laboratory itself. So it came about that pioneers of such important physical techniques as Raman spectroscopy, heat capacity and magneto-optics found themselves exercising their skills and insights on the subject matter of inorganic chemistry—to the enormous long-term benefit of the latter.

In creating a third Chair in Chemistry, the University brought to an end the process by which two Chairs precessed among three departments (the analogy with the game of Musical Chairs was not lost on the academic audience). The new Professor, the eminent solid-state chemist Stuart Anderson, wanted old and new to cohere as amicably as possible and, to that end, decided to institute a common room where members of the academic staff could meet. The formidable organising (if not aesthetic) talents of the Departmental Secretary Ann Wallace were engaged to choose various chairs and banquettes. To begin with, these were installed in a room on the third floor, then newly constructed on top of the massive stone structure from the 1950s. (Fortunately, the mighty thickness of the walls, while making it more or less impossible to rearrange the internal spaces, at least made it easy to support a whole new floor of offices and laboratories on the top). However, the third floor room soon became too small to accommodate the increasing numbers, in those distant days of expanding numbers in teaching and research. So Anderson had an inspired idea: why not use the old Abbot's Kitchen?

The central benches were ripped out; the fume cupboards became showcases for historic pieces of apparatus. The walls were hung with venerable group photographs of research students and their mentors, discovered stacked away in the basement. In place of the benches, tables and chairs from the anonymous third floor room were installed and the Abbot's Kitchen was launched on its new lease of life as a forum for chemical scientists to congregate.

Twice a day the seductive lure of coffee (and that most English of beverages, tea) brought faculty members and graduate students together—but to what purpose? Does the quantifying imperative that rules every facet of publicly funded academic life in early twenty-first century Britain seek return from those who occupy this space each morning and afternoon? If so, how many pounds per square metre is the conversation invoiced at? And even before you get to that part, how can you value it? Certainly the octagonal shape, high ceiling and charming atmosphere combine to make the space admirably suited for its new purpose. Informal and un-programmed chat between student and supervisor, or experienced postdoc and newcomer, also do much to smooth the progress of a project and negotiate its pitfalls. Even if the topic of conversation has nothing to do with chemical (or any other) science, exchanging a few civil words over coffee does make it easier to borrow a piece of equipment from the room next door or pick up some previously unsuspected fact. May it not be that the Abbot's Kitchen is at least as productive of new knowledge now as when it was filled only with the fumes of nineteenth century chemical reactions?

By dwelling on these examples of meeting rooms in London and Oxford, it may be that I have left an impression that 'coffee and conversation' is a peculiarly British catalyst for scientific progress. That would be quite mistaken, as my own personal experience, at least of mainland Europe, testifies. As I shall tell later on, the USA and Japan are entirely another story. Of course, Europe is a word encompassing quite astonishingly diverse cultures and social habits so at first sight it might seem odd that underneath all this kaleidoscope of behaviour and assumptions, there are plenty of instincts that we hold in common. Among these is a love of gossip and argument, not a bad basis for getting together in collective enterprises, though one that can easily lead to misunderstandings along the way. When the collective enterprise in question is a scientific one, you might think that at least the goal would be fairly straightforward to define and agree. Not so, as plenty of instances confirm. However, what I am trying to do here is not to describe the grounds for debates of that kind which I have taken part in, as to sketch a few of the places where the wheels of debate have been oiled (if that is the right word) by coffee and converse.

In my own personal scientific (and geographical) odyssey, a small village just beyond the outer suburbs of Geneva holds a special place. The village, called Cologny, is best known nowadays for a stratospherically expensive restaurant frequented by bankers, international diplomats, film stars and the like. On a hill sloping down towards the lake, the village commands a spectacular view across the water towards the city centre and the line of the Jura mountains to the north, crowned with snow for six months of the year. Half a mile further up the hill beyond the village square, however, on the road leading to the next village, La Capite, and the French border, lies another reason for the name Cologny to become known, at least in the scientific world, far beyond the frontiers of Switzerland.

Climbing the hill, the view opens out in an even more spectacular way, now to the south as well as the north, revealing a distant ethereal prospect of Mont Blanc some fifty miles away. Not surprisingly, this area became popular long ago with the well-heeled citizens of Geneva escaping the city, and rich foreigners seeking refuge in the benign fiscal air of Switzerland. Numerous estates were carved out of the surrounding farmland, one of the earliest being a fine Gothic revival edifice that would not have looked out of place in Victorian North Oxford. Two hundred yards beyond on the other side of the road, an impressive pair of wrought iron gates flanked by stone pillars opens on to a sweeping driveway leading up to another substantial villa, this time with a mansard roof in the French classical manner. Closer inspection, however, will reveal that this imposing (not to say intimidating) structure has two further extensions placed symmetrically on either side, but in an altogether more mundane, functional and, indeed, somewhat banal 'international' style. In the decade from the late 1950s to the end of the 1960s, these latter housed chemical laboratories for, at the time, this was the home of the Cyanamid European Research Institute, otherwise known as CERI.

More about the huge success of CERI in several quite distinct areas of chemical research can be found elsewhere but, for the immediate purpose of describing 'conversation rooms', we have to look, not at the original

Fig. 18. The Cyanamid European Research Institute, Cologny, near Geneva, Switzerland.

classical villa or the flanking laboratories, but towards a much smaller building just inside the gateway on the right-hand side. Its architectural style was a miniature version of the main villa and, as originally conceived, it must have housed a porter or housekeeper. When the last banker owner moved out of the main house and the American Cyanamid Company took over the site, the gatehouse became a small cafeteria where the scientific and secretarial staff took coffee, lunch and afternoon tea (the citron for me). So another conversation room was born.

In my opinion there are two reasons why this small space was so successful in fomenting novel ideas and perspectives—apart (that is) from the abilities and creative spirit of the people who congregated there and the quality of the cooking that drew them. One was simply the location. Being a few miles outside the city, and with only one infrequent bus service passing the door, it was just not convenient for many to go home at lunchtime, as was often the case at that time in continental Europe. So as well as the home cooking, ideas and new results were chewed over by a large quorum of the CERI scientists.

The second reason, reinforcing the first, goes back to the way the Institute had been set up and organised. The Cyanamid Company had decided to hire up to six senior scientists, each of acknowledged eminence in their own fields, which were chosen as ones the Research Director and his advisors thought might have some relevance to the business over the long term. For example, thinking about pigments and phosphors, they employed Christian Klixbull Jorgensen, then famous worldwide as an inorganic chemical spectroscopist. Others included Manny Moser (physics of new compound semiconductors), Bob Hudson (organo-phosphorus and -arsenic chemistry) and Fausto Calderazzo (organo-metallic chemistry and homogeneous catalysis). The seeds of fruitful cross-fertilisation between disciplines, outside the traditional boundaries of organic, inorganic and physical chemistry (or even between chemistry and physics) were therefore planted from the outset. Klixbull Jorgensen, who was trained as a chemist, once jocularly remarked to me that he was really an inorganic physicist.

However, there was a final ingredient crucial to the success of this multi-disciplinary mix, and to this day I have no idea whether it was accidental, the product of subtle thought by the managers of the Cyanamid Corporation's Central Research Division, or just an outcome of restricted budgets. Quite simply, it was the fact that all the groups working in these disparate fields were very small; only three permanent staff scientists and one technician, plus from time to time a temporary outsider (of which, in 1962, I was one). If we were not to get increasingly fossilised in our own individual specialities, we had to talk to one another. So lunch sometimes stretched out well beyond the hour allotted by the management as debate raged and the Formica-topped tables became progressively covered in pencilled formulae and diagrams.

Cafeterias are natural foci for discussions: coffee and conversation go together inexorably. A further and, if that is possible, even more eclectic example can be found some 100 miles south of Geneva in the French city of Grenoble, at the Institut Laue-Langevin, the European centre for high-flux neutron scattering. The origins and nature of this admirable institution are described in more detail in Chapter 3. Suffice it to say here that beams of neutrons are used to solve problems across all the sectors of condensed matter science, ranging from basic physics through materials science to solid-state chemistry and biology, not forgetting engineering. Because neutrons come from either nuclear reactors or particle accelerators, beams of these useful particles are not to be found in university laboratories, so a suite of instruments exploiting the various characteristics of wavelength, polarisation and so on have to be brought together round the neutron source itself. Putting together such an assembly is expensive, so the establishments housing them have to be organised on a national (or, in the case of the ILL, international) basis.

Herein we have the seeds of an astonishingly diverse enterprise—a multi-disciplinary laboratory that is also multi-national. This diversity manifests itself at every level, from the Board of Directors (English, French, German and—in my time as Director—drawn from nuclear physics, theoretical soft condensed matter physics and inorganic chemistry) to the expert panels who select the experiments to be performed. The latter are ten in number, reflecting all the sciences from particle physics to biology, each one being made up of representatives from all the member countries. Above all, though, it is the hoards of scientists from all these countries who throng the halls and corridors of the Institut on the way to do their experiments. Since the average length of time that any scientist has access to an instrument for one experiment is only a few days, the crowd is dense and ever-changing.

Where do they find their focus? You guessed it—the cafeteria. Indeed, where else would a physicist from Germany meet a biologist from the UK and hatch a joint project? Not only is this kind of so-called 'facility-based' science a strong boost to inter-disciplinarity, but the facility itself creates the focal point—and the solvent is coffee.

To conclude, are there any generalities to be extracted from this small, unscientific and entirely personal sampling of milieux where scientists have shared their ideas and results? All the institutions I have mentioned here can be characterised as extremely productive (and indeed creative) in their output of new knowledge; I feel lucky and proud to have been associated with them. But did the coffee rooms I have described contribute to that, or are they just a pleasant by-product or even, at worst, a distraction from more serious business? In other words, might not these establishments have been just as productive if the rooms had been used for some less frivolous purpose? No professional social scientist worth their reputation would want to draw any correlations from such a narrow base, but let me at least try.

First, the last question, by way of a counter-example. Communal coffee-drinking is much less common in laboratories in the USA, especially in industrial laboratories, where a high seriousness about spending the company's time and money tends to rule. Not that coffee is not drunk—it is, throughout the day and in industrial quantities—what is lacking is the communal assembly. Many years ago, I spent a few months at the research centre of a major US corporation and found that, while a coffee trolley was parked each morning in the corridor outside the laboratories and offices, everyone came out of their rooms, filled their mugs and went straight back in again. In parallel with that, once a week there was a formally convened group meeting lasting at least an hour, at which most of the business was pretty mundane (timetabling experiments, planning equipment deliveries and so on). After a while, I was driven to observe that if, instead of returning to their desk or bench, the coffee drinkers had lingered a few moments in the corridor, practically all of this business would have been covered by what physicists call 'pair-wise interactions'. My words fell on deaf ears. I do genuinely believe that the laboratory was not only less efficient, but also less productive, as a result. In Japan, too, sitting around in apparently aimless chatter is not part of the national psyche. Nevertheless, a sign that the Japanese themselves have become aware of the benefits of un-programmed debate is the fact that, over recent years, quite a few new laboratory buildings of corporations and universities having been designed to contain 'discussion zones', set aside with easy chairs. For myself, however, I have not observed them being used very intensively.

Given that both the USA and Japan are world scientific powerhouses, what is it that they can possibly be missing? Volume of output and also productivity are certainly not lacking. Insight and creativity are harder to estimate but in many fields, Europe can stand proud, especially in relation to the means at its disposal. May not the culture of 'coffee and conversation' have something to do with that? Trying out partly formed ideas or voicing uncertainties lubricates the path to innovation and where better than in the kinds of environment I have been describing. So let us drink a toast (in coffee) to the conversation rooms of Europe and the debates they have hosted.

chapter 3

The Institut Laue-Langevin: A Crucible of European Sciences

"May I welcome you to this lecture and thank you for coming in such large numbers."

In the first minute of my first Friday Evening Discourse at the Royal Institution I broke two house rules: I started by greeting the audience, and I began to address them in what I hoped was at least an approximation to a foreign tongue. I hope they excused the first misdemeanour since a greeting was in order on such an occasion. The second had a more mundane and practical purpose: to give a brief flavour of the way that I had been required to communicate from time to time in public in my years as Director of the Institut Laue-Langevin, and to indicate that not all communication in Europe proceeds in English. There is a cohesive Europe already in science, long pre-dating Maastricht, Nice and other Treaties enlarging and re-ordering the European Union, and the United Kingdom is indissolubly part of it. My purpose here is to explain how this cohesion was built up in one field of research (neutron scattering) and how, from its early beginnings in an apparently esoteric technique, it expanded to embrace nearly all the sciences, and an extremely large number of participants. To understand the present, we need to go back somewhat into the past, to indulge in what the green Michelin Guides to the French regions call '*un peu d'histoire*'.

After our time travel, we can then take a short tour through Europe, down to Dauphiné, in the Rhone-Alpes region of France.

For many years now, scientists who study the largest and the smallest objects in the Universe have had to use more and more elaborate and expensive tools, to the point where no one country could afford them. Equipment such as radio telescopes and high energy particle accelerators is now made available through international collaborative agreements. The form of such agreements varies greatly, from the fully-fledged long-term organisations such as the Centre Europeen de la Recherche Nucleaire (CERN) to *ad hoc* arrangements for access to existing national facilities. Naturally, the development and siting of such facilities is a highly political matter, and national governments take a strong interest, as witness the inauguration of the Electron Positron Collider (LEP) at CERN in 1990, which was attended by the Presidents of France and Switzerland, the King of Sweden and the Crown Princess of the Netherlands among others.

In contrast, chemists, biologists and condensed matter physicists use much smaller and simpler apparatus. For example, to make crystals of molecular superconductors (a topic of great interest in my own research), quite ordinary laboratory glassware and an electrochemical cell are all that is required. However, even a simple chemist must try to find out what he has made, first in terms of composition, by chemical analysis, and next the crystal structure. The first method that comes to mind for solid state structure determination is, of course, X-ray diffraction, a technique in which the Royal Institution has a long and distinguished reputation. Still, though X-ray generators and diffractometers have become much more complicated and costly than in the Braggs' day, it remains common for them to be found in research laboratories. For example, the Royal Institution has two in the basement.

Suppose, however, that the crystal whose structure we wish to determine contains light atoms as well as heavy ones. For instance the molecular charge transfer salts just referred to, that behave as superconductors, consist of cations which are aromatic heterocycles made from carbon, hydrogen and sulphur, and anions containing heavy metal atoms. Or (as is sometimes the case in the same class of compounds), the crystal contains unpaired electrons that interact with one another on neighbouring atoms to form a superstructure. Then X-rays are less useful, and we need a more subtle probe. This is the point at which the small sciences start to impinge on the larger ones, because the probe in question is the neutron, and neutrons come from nuclear reactions.

Neutrons are very useful particles for studying condensed matter. Being electrically neutral they interact relatively weakly with the sample, and therefore constitute quite a 'gentle' probe compared with electrons or X-rays. The interaction being with the atomic nuclei in the sample (rather than the electronic shells, as with the latter two probes), the scattering probability varies from atom to atom across the Periodic Table in what appears at first sight to be a random fashion, but which is determined in fact by the nuclear structure. That is why, in striking contrast to X-rays and electrons, the intensity of the scattering of neutrons by hydrogen atoms is greater than by gold atoms. Another advantage of the neutron is its magnetic moment. The magnetic moments generated by unpaired electrons in the sample therefore interact with the neutron beam to produce scattering that, in the case of long range order, manifests itself as extra diffraction peaks. Analysis of such peaks tells us the orientation of the atomic moments in the crystal, and whether they form a ferromagnetic or anti-ferromagnetic superlattice. Finally, unlike photons, neutrons have mass, which means that they carry momentum. Not only, therefore, can they be scattered elastically, with the resultant wave-packets interfering to create diffracted beams from periodic structures, but they can also be scattered inelastically by transferring energy (and momentum) to or from the sample.

Neutron scattering can thus provide valuable information about vibrational or magnetic excitations having energies up to about 200 meV or 1,600 cm^{-1}.

Nearly all neutron scattering instruments designed to work with a steady continuous incoming beam of neutrons contain the same basic elements. There is a monochromator to define the wavelength (and hence velocity and momentum) of the neutrons that impinge on the sample. If the sample is a crystal it can be rotated in the collimated beam. Finally a detector that can rotate around the sample collects the scattered neutrons and converts them into an electrical signal that can be amplified and displayed. A whimsical view of such a mechanism is shown in Fig. 19. The reality is a bit more complicated, two examples at the Institut Laue-Langevin (ILL) being given in Fig. 20.

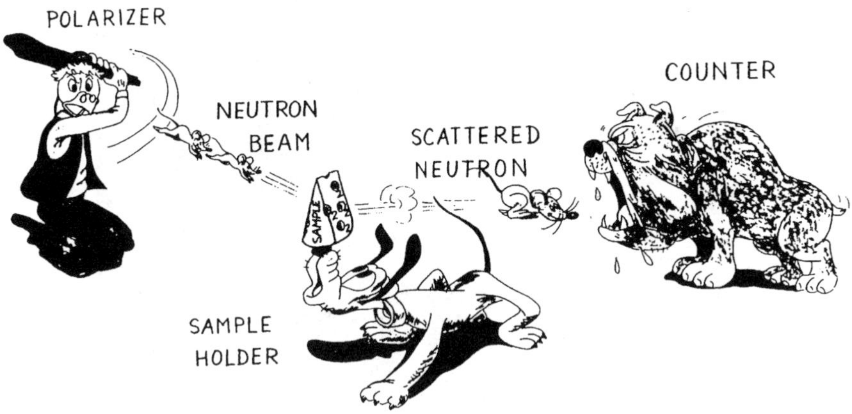

Fig. 19. Schematic view of a typical (?) neutron scattering experiment (with thanks to Enrico Rastelli, Parma).

The origin of the neutrons that feed the instruments in Fig. 20 is in a nuclear reactor. Nuclear reactors work by creating a self-sustaining chain reaction in which ^{235}U nuclei divide to give fission products and neutrons, which are slowed by a moderator (graphite in the earliest designs, heavy water in more recent ones). The slow neutrons are then captured by other ^{235}U nuclei to release even more neutrons, and also heat which can be harnessed to make steam, driving a turbine and a dynamo to make electricity.

The first generation of research reactors were built after the Second World War to investigate the physics of the fission process, the design of reactors themselves, and the behaviour of the materials used to make them when they are subjected to intense radiation for long periods. The reactors in question were, of course, all owned and operated by national governments and located in Atomic Energy Research Centres, especially in Britain (Harwell), France (Saclay and Grenoble) and the USA (Argonne, Oak Ridge). It was only in the 1950s that there was sufficient pressure of interest from a wider scientific community to make it worthwhile to make holes in the sides of these reactors to let beams of neutrons out.

The administrative arrangements that were set up in these three countries to help condensed matter scientists to exploit the neutron beams have had a strong influence on the way that neutron scattering techniques developed. In my opinion this piece of history shows the UK in a very good

(a)

(b)

Fig. 20. Typical instruments installed at the Institut Laue-Langevin (ILL). (a) The multi-detector diffractometer D20, (b) the neutron spin-echo spectrometer IN15 for low energy inelastic scattering. Note how the parts correspond to those in Fig. 19.

light, and reflects great credit on those who took the initiatives. In essence, the Research Reactor Centres run by the government agencies in America (Department of Energy) and France (Commissariat d'Energie Atomique (CEA)) were used almost exclusively by scientists from the Centres themselves. By contrast in Britain, University scientists quickly became associated with the work, culminating in an arrangement whereby the Department of Scientific and Industrial Research, later Science Research Council, still later Science and Engineering Research Council (SERC) rented up to half of all the beam time on the Harwell reactors from the UK Atomic Energy Authority on behalf of the university community. As a result, the range of science being done with neutrons widened greatly in Britain, especially towards chemistry and biology. Not that classical solid state physics was left out, and there grew up a flourishing national school, especially of magnetism. The sheer numbers of people involved were also impressive, up to nearly 600, drawn from nearly every University in Britain, a state of affairs that continues to this day.

Why then should I be talking about a research reactor in Dauphine and not, say, Oxfordshire? The next chapter in this history was a decision to have reactors devoted exclusively to producing neutrons, and that is the part where science starts to interact with politics. In the early 1960s British scientists tried to persuade their government to build a High Flux Reactor for their use, but they were not successful. In parallel, on the continent of Europe, the *rapprochement* between France and Germany, impelled by the political alliance of Presidents de Gaulle and Adenauer, was in need of symbols: a research reactor dedicated to basic science for peaceful purposes was an attractive project. Thus the idea of the Institut Laue-Langevin (ILL) was born, named after an eminent German physicist, Max Von Laue (who invented an early form of X-ray diffraction at the same time as the Braggs) and Paul Langevin, the French physicist who did significant early work in magnetism.

Two years later, a new campaign was mounted in Britain to persuade the government to build a comparable facility here, but the final decision was to join in with the French and Germans in ILL. However, by that time the ILL was already six years into its life as an institution, though only two years from the start up of the reactor. In my opinion that delay in joining meant that when the UK arrived on the scene, not only had the decision on siting been taken long since, but many technical and administrative decisions too.

In a financial sense we became full partners, paying our share of the capital as well as the running costs. But we were joining a going concern on the continent of Europe, which we had taken no part in formulating—does that perhaps remind you of something else?

The first decision of course, had concerned the siting of the Institut. Given the lack of enthusiasm in the Federal Republic for matters nuclear, balanced by great enthusiasm in France, where the CEA occupied a pivotal position in national policy, it was clear that the reactor should be in France. But where? Now for some historical geography.

In South-western France, in the region now called Rhone-Alpes, the town of Grenoble had always occupied a strategic position, situated as it is close to the border of Dauphine and Savoie. The latter became part of France only in 1860 so the frontier area had been heavily defended since the seventeenth century. Most conspicuous of the monuments remaining from this era are the numerous forts constructed by Vauban, including the Forteresse de la Bastille overlooking Grenoble. The town itself had been surrounded by fortifications and army barracks, whose regular geometrical disposition earned the western part of the town the name of 'Polygone'. Part of Grenoble's strategic importance lay in the fact that two major river valleys meet, and the Isère and Drac join together just beyond the outer ring of fortifications, creating a long peninsula in the flat bottomed glacial valley which served for many years as an artillery firing range. When, following the Second World War, the CEA was evaluating potential sites for its Centre d'Études Nucleaires (CEN) the existence of such a large block of land owned already by the Government, close to a town that contained a distinguished scientific and technical university, made Grenoble a prime candidate. It is instructive to note that a very similar combination of considerations led the UK Atomic Energy Authority to establish its Research Laboratory on the site of an airfield at Harwell, within easy reach of Oxford.

After the CEN(G) was set up on the Polygone site, other national laboratories, such as those of the Centre Nationale de la Recherche Scientifique (CNRS) followed. As discussion about potential sites for the ILL got under way in the mid-1960s, what had become known as the Polygone Scientifique in Grenoble was a clear favourite, rendered even more attractive by the offer of the CEA to lease a sizeable, and hitherto unexploited, piece of the peninsula to the ILL Associates at a peppercorn rent.

So the site was ideal. But one should also not neglect the influence of powerful individuals on the course of history. The individual in this case was Louis Neel, Physics Nobel Prizewinner of 1957, Professor at the University of Grenoble and eventually Rector of the University. His field was magnetism, in particular the behaviour of ferromagnets and antiferromagnets, to study which neutron diffraction had already proved a most effective tool. The other main protagonist was Professor Maier-Leibnitz of Munich, a pioneer in developing instruments and techniques for neutron scattering. Portrait photographs of these two founding fathers quite rightly adorn the corridor of the ILL. Furthermore, in 1989 the CEN(G) and CNRS agreed to rename the complex of laboratories on the peninsula between the rivers Isère and Drac as the 'Polygone Scientifique Louis Neel', at a ceremony attended by the gentleman himself. Earlier, in 1987, at a less official and more lighthearted occasion, celebrating the 20th anniversary of the signing of the original ILL Convention, the ILL staff conferred on Professor Maier-Leibnitz the title of 'Grand Maître de l'Ordre de la Confrerèrie du Taste (not Vin—though it flowed—but) Neutron'. The ceremony is illustrated in Fig. 21.

So what was this ILL that was set up on 1967? The question may be answered in two ways, by examining in turn what may be called the 'hardware' (by which I mean the reactor, instruments etc.) and then the all-important 'software' (namely the people and procedures). Let us look at the hardware first. In the aerial view of the Institut in Fig. 22, the cylindrical building at the center, behind the ESRF in the foreground, houses the neutron source, called the Reacteur Haut Flux (RHF).

The reactor is special because it was designed from the outset to make neutrons, not electricity or heat. At its centre is a single fuel element containing about 10 kg of 93% ^{235}U (HEU) in the form of U-Al alloy in plates, of which five are used each year. The heat that comes from the fission process is taken away by heavy water (D_2O) forced between the plates and a secondary heat exchanger transfers the heat to water from the river Drac. The power dissipated by the RHF is about 58 MW, heating the river about 0.1 C. A cross-section of the RHF appears in Fig. 23. The heavy water tank is surrounded by light water, then concrete, and finally the reactor dome. All operations of putting in and taking out the fuel element take place in the top half, and experiments in the lower half.

Fig. 21. Professor Maier-Leibnitz, one of the founders of the ILL, being honoured at the party celebrating the 20th anniversary of the foundation.

The number of instruments that can be gathered around the reactor is limited by the space available, though the hall is large. An important innovation, since widely copied at research reactors throughout the world, is to conduct the neutrons away from the reactor through pipes, which operate like wave-guides by total internal reflection of the neutrons from nickel coated internal surfaces. As one can see from the schematic layout

Fig. 22. An aerial view of the ILL on the peninsula between the rivers Isère and Drac, with the Bastille and the Belledonne mountains beyond. The large ring in the foreground is the European Synchrotron Radiation Facility (ESRF).

in Fig. 24, about half of all the instruments are placed on neutron guides, some of which are up to 120 metres long. Along with guides, another new development especially associated with the ILL is the widespread exploitation of neutron polarisation, most readily accomplished by Bragg reflection of the neutrons from a magnetised crystal.

In all, some 30 instruments that deal with powders, crystals, liquids, surfaces etc., are available to scientists from the member countries to come and use. Additionally, there are long-term experiments of the kind set up by particle physicists, who want to use the neutron itself as a laboratory, and examine its structure and properties. Examples of the latter include a classic long-running experiment to determine the lifetime of the free neutron (actually about 17 minutes) using very slow (otherwise known as ultra-cold) neutrons trapped in a bottle. Furthermore, many attempts have been

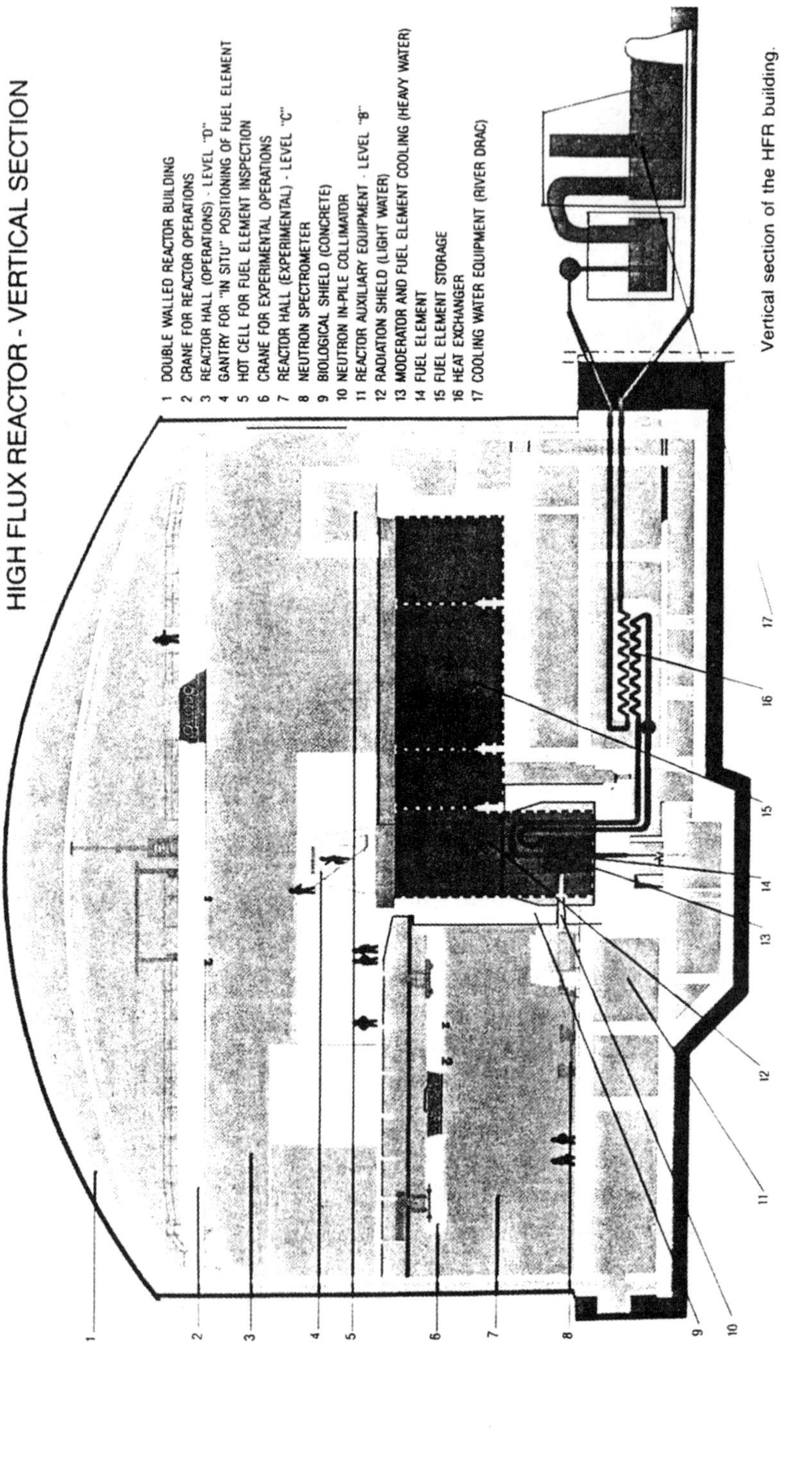

Fig. 23. A cross-section through the ILL reactor and building.

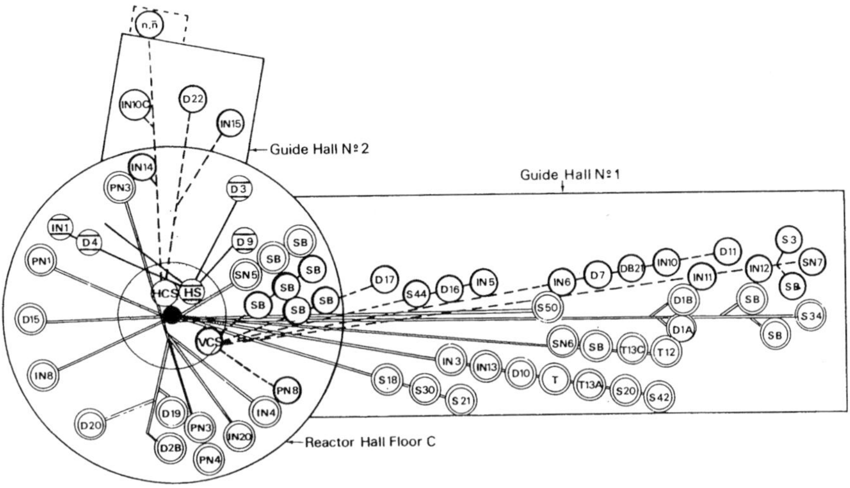

Fig. 24. Layout of the neutron-scattering instruments around the reactor and in the guide-hall at ILL. Each circle represents an instrument.

made to determine if the neutron has a measurable electric dipole moment. If so, it is very small (less than 10^{-26}e.A), but any non-zero value would provide a severe test of what high energy particle physicists call the 'Standard Model'. As large accelerators progressively price themselves out of the market in particle physics, precise measurements on composite particles like the neutron and proton will increasingly prove cost-effective substitutes.

As far as other branches of science are concerned, neutron scattering has penetrated them all. A few highlights only will serve to illustrate the rich diversity of topics to which neutrons have contributed. When high temperature superconductivity was discovered in 1987, the phases exhibiting this remarkable phenomenon were oxides containing heavy metal atoms like lanthanum and strontium. Neutron diffraction was therefore a much more convenient tool than X-ray diffraction to determine their crystal structures, and in particular small displacements of the lighter oxygen atoms from their ideal positions in the lattice. Figure 25 shows such a structure, determined at the ILL.

From an unrelated field, but of equal significance, the photosynthetic reaction centre (PRC) is one of the most important molecular complexes in biology because it is the entity which converts the energy of sunlight,

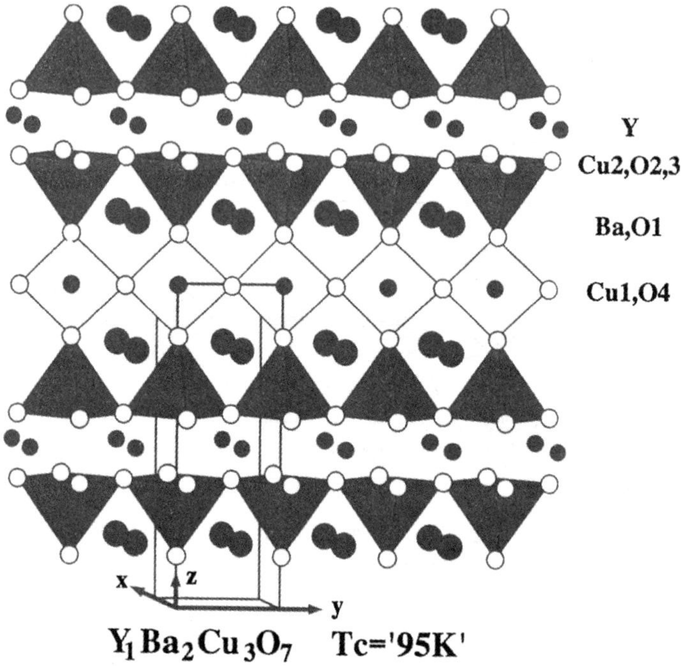

Fig. 25. The crystal structure of the high temperature superconductor $YBa_2Cu_3O_7$ as revealed by powder neutron diffraction at ILL.

captured by chlorophyll 'antennae', into chemical energy by initiating the separation of a proton and an electron. The PRC is a membrane protein whose fragile tertiary structure is held in place by the lipid molecules constituting the cell wall membrane around it: remove the lipid and the protein falls apart. Although the crystal structure of the PRC was determined by X-ray diffraction in 1986, an achievement that won the Nobel Prize for Chemistry for Professors Huber and Diesenhofer, they could not locate the all important molecules of lipid because they are very strongly disordered. Exploiting the vast difference in scattering intensity between H and D, neutron diffraction solves the difficulty because membranes can be prepared to contain hydrogenated protein and deuteriated lipid, or *vice versa*.

Outside the field of pure science, neutrons are making increasingly important contributions to the applied sciences and engineering. A striking example is the determination of residual stress in engineering components.

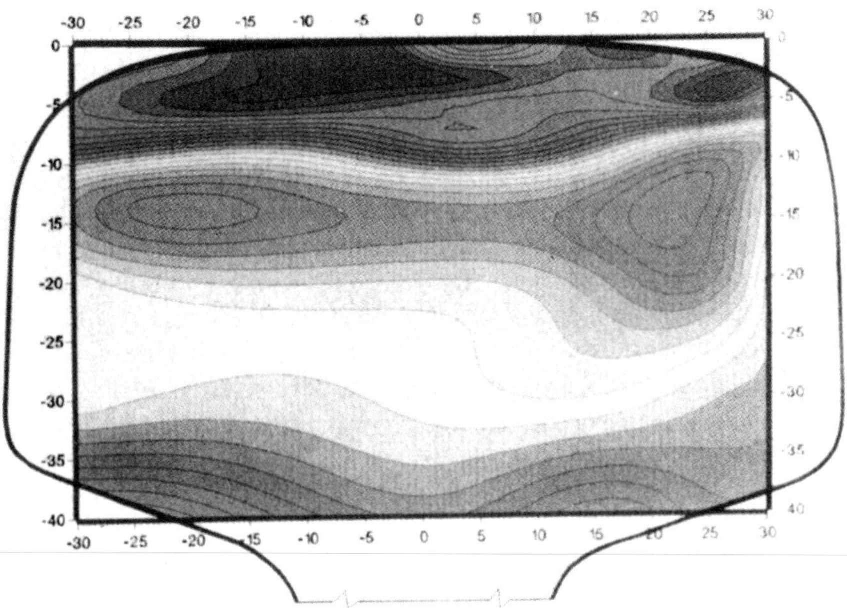

Fig. 26. Contours of residual stress inside the head of a railway line revealed by neutron diffraction.

Because neutrons interact relatively weakly with condensed matter they can be used to 'see through' quite large objects. Until recently the only way to measure the residual stress inside a large piece of metal was to bore a small hole in it and insert a strain gauge—but of course that changes the quantity being measured! By carefully measuring the angle of reflection of a well collimated monochromatic neutron beam from one or more lattice planes in a solid specimen, it is possible to determine their spacing, and hence the residual stress, at different points throughout the specimen. An example (actually a piece of railway line) is shown in Fig. 26.

A specially interesting feature of the way in which the 30 'scheduled' instruments are used is to perform experiments in quite disparate fields of science using the same instruments. An example from my own personal experience illustrates the point. The diffractometer D16 was designed for studying the texture of fibres, especially of biological molecules such as DNA and collagen, because it was equipped with a multi-detector. We pointed out that by adding a cryostat to reach low temperatures, the same set-up could be exploited to examine the incommensurate magnetic structures of crystals, a topic of great interest in solid state physics. Such multi-disciplinarity often leads to unexpected cross-fertilisation between disciplines: the physicist who devised the modelling routines for the magnetic structure work later worked on the application of multidetectors in biological Laue-diffraction. The widespread use of Small Angle Scattering (SANS) instruments provides another illustration (now there are three at ILL). The degree of orientation in liquid crystal polymers, the structure of a gene-repressor complex and diffraction from the flux lattice in a high T_c superconductor, are just three topics examined by this versatile method.

So much for the hardware (reactor, instruments, data): in some ways even more fascinating is the 'software': how is all this diverse activity organized? How does the ILL function both as a community in itself and as a node within an international, even global, community? My starting point for answering these questions may appear at first sight to be rather prosaic: the legal framework. Yet in truth it has a powerful influence on the way the Institute operates. ILL is a French company, a *Societé Civile*, with four shareholders. That is quite important: it is not a supra-national body like CERN or the European Molecular Biology Laboratory (EMBL). It (and its employees) therefore exist within the French legal system and that goes for the reactor, which is subject to the rules of the '*Service Centrale de Sureté des Installations Nucleaires*' (SCSIN), then a branch of the Ministry for Industry in France. It goes, too, for the tax system, the working conditions, and the staff representation through the French Unions. Those who have not had the pleasure of negotiating with a group of '*Representants Syndicaux*', including CGT, CFDT and FO or of being interviewed by the Head of the SCSIN (shades of the Headmaster's study!) have truly missed some of the more piquant moments along the road towards building a new Europe.

The shareholders, too, can sometimes be less than accommodating to the aspirations of the company's 'Chief Executive'. At first sight an arrangement of three equal partners looks to have the in-built stability of a three-legged stool. (I spent a good deal of energy arguing as much to UK officials during Britain's periodic efforts to reduce its payments.) The sad truth though is that budget discussion in these circumstances resembles a Dutch auction: the final figure must be agreed by all parties, but often at the level proposed by the most indigent (not always the UK). The shareholders are represented by a Steering Committee, not at all to be confused with the Institute's Scientific Council. During my time as Director I initiated a system whereby an account of what (in my view) had been the most noteworthy new science was pressed on the members of the Steering Committee: I had the feeling that the level of interest was sometimes not high, though I insisted on subjecting them to it. Budgets, managerial control, and manning levels are the topics most discussed by the representatives of the funding agencies. It is unfortunate that they are often less interested by output than by input.

The twice yearly Scientific Council is the real meeting of scientific minds: in addition to the Council itself, there are no fewer than eight subcommittees covering everything from nuclear physics to biology, manned by experts (colleagues, friends, rivals (?)) who come together in Grenoble twice a year to decide which of the proposed experiments should get precious beam time that is 2:1–3:1 oversubscribed. Sitting in these groups, almost never does one hear a nationalistic argument in favour of a proposal: scientific novelty is what counts—this is where the unity of European scientific endeavour shows itself most potently. Social encounters form an important part of these occasions too; a good dinner for the 80 odd members of the subcommittees, lots of talk, in-jokes ('in', that is, to ILL hands, not just from one university or country). This is truly the mechanism that smoothes the scientific wheels.

Scientific Council and Steering Committee meetings punctuate the seasons, like sowing and harvest, but what about the trivial round and common task in between? The ILL is truly a factory for science. In a full year of 5 reactor cycles, 850 scheduled experiments are performed, yielding 550 publications. The mean length of an ILL experiment is 4.5 days, though some take 2 weeks and others, for example SANS, 12 hours. Consider the implication of such a densely packed schedule for the infrastructure of the Institute: at the beginning of each experiment a new team arrives; perhaps a knowledgeable professor, perhaps a nervous new graduate student. Everything must be operational in readiness for them: the reactor, the apparatus, detectors, computer, cryostat, etc. A scientist who has prepared a sample and shipped it 1,000 km for an experiment lasting 48 hours, in the knowledge that he will not get beam time for another 6 months gets very upset if time is lost because of a blocked syphon on a cryostat or a dead board in the computer.

That is why, out of 480 people on the staff of ILL more than 200 are in the technical side—the show must go on.

And in the corridors too, the scientific contacts continue. Folklore says that if you stand on the corner of Piccadilly Circus long enough, you will meet everyone you know. In the world of science, I have long had the impression that the entrance hall of the ILL holds the same characteristics: solid state physicists from Berlin greet chemists from Zaragoza, and biologists from Heidelberg compare notes with polymer physicists from London. Furthermore, the 'crucible' of the title above is not just ILL itself but within it an even more potent catalytic centre: the coffee room! How many new scientific collaborations started there?

So much for the fortunate experimentalists who come to use the ILL equipment and cross-fertilize their ideas. What of the denizens who serve this shifting population? At the apex of the organisation lies a small culturally mixed group, the Directorate, consisting of one English, one French and one German scientist, each nominated by their national representatives for a fixed term. However, in the remaining staff complement of 480, the three principal participating nations are by no means equally represented. Herein lies one of the biggest long-term problems of UK involvement in international scientific enterprises like ILL: how many young scientists, technicians, and secretaries want to go abroad and take part? With the enthusiastic help of colleagues in the Grenoble Universities, ILL and the neighbouring European Synchrotron Radiation Faculty (ESRF) started a residential course of lectures and demonstrations for young graduate students, on neutron scattering and synchrotron X-ray studies—how to use these large instruments. At the first course in 1991 there was a single British student, despite all our advertising. One year later, there were four out of a total of 63. By comparison, Spain sent seven.

A similar trend is found in appointments to scientific posts at the Institute. The number of scientists from the UK with permanent contracts at ILL is equal to the number from Australia. Staff scientists are recruited largely from those holding temporary contracts and there is no quota among nationalities. My duty, with the advice of the other Directors and Senior Scientists, was to appoint those perceived to be the best. The problem is not quality, but that our young scientists are not entering the competition: the French and Germans are. A photograph of the ILL football team of the 1980s shows a good distribution of nationalities (French, Germany, UK, others): why not scientists? The message must be that collaborative European science is an exciting, exhilarating, and sometimes baffling arena, but the UK must be part of it, and that means fully part, not semi-spectators. We must also be aware that far from joining a world that is foreign, the same world is foreign to everyone else, and we can take part in moulding it as well as others.

The ILL Scientific Secretariat unwittingly provided a splendid metaphor on the nature of the Institute when they defined the rules for drawing up the statistics about the subcommittees' distribution of beam time between the participating centres and others. A Franco-German or Anglo-Swiss experiment is easily accommodated but how should one treat an experiment proposed by a team consisting, say, of a British scientist and a German member of the ILL staff? Is it Anglo-German for statistical purposes? No: all members of ILL staff are counted as one category, irrespective of their country of origin. So the conclusion is clear: ILL is a 'country'! It is not a piece of France or Britain or Germany but *sui generis*, like Andorra or the Vatican State.

What is being constructed at establishments such as ILL is a way of conducting research that transcends national boundaries, because facilities are provided beyond the means of individual states. That further means that where such 'central facilities' are concerned, national funding agencies should give higher priority to international solutions so that the facilities are truly 'central'. It remains an unfortunate fact that when funds are tight (as in practice they always are) there is a tendency for national funding agencies to look to the interests of their own establishments first. On more than one occasion I was drawn to compare ILL, seen from the viewpoint of the British scientific mandarins, as the 'far away country of which we know little'. In return, international establishments like ILL need to be tied more closely to the national laboratories, by regular exchanges of personnel and equipment. At ILL such contact has always existed from the French and German sides, but much less so with the UK national laboratories. The benefits of increased cross-fertilisation would be large, and mutual.

Following from that, it is of the utmost importance that commitments are entered into wholeheartedly, and on a long-term basis. The UK has an unenviable reputation of grudging and backsliding in its contractual relations with the major European scientific collaborations. Of course, we have to think long and hard before deciding whether to join a project, but having thought, we should not constantly be rethinking. And when we are doing the thinking, the group whose views must be sought, and whose opinions are paramount, is that of the scientists themselves. International science is a beautiful example of an Adam Smith market: scientists go where they can get the best results. They will beg, borrow or steal travel funds, and beat on the doors of the Institutions that provide the best facilities.

And what is the future of ILL in this context? In the early 1990s SERC (as it was then called) wanted to lower the UK contribution, while the reactor was also shut down for a major overhaul. Now, in my opinion, the future is extremely bright. Discussions about future European neutron sources (steady state and pulsed, reactor and accelerator) are underway. Now overhauled, ILL has a new life expectancy of at least 15 years. Therefore it will be the 'next' European reactor source. But 'European' has to mean just that: not Franco-German with the UK playing a minor role.

Already there are three scientific members at ILL (Spain, Switzerland and Austria) in addition to the 'Big Three' of Britain, France and Germany. I had the pleasure of signing the contract with Austria. I also spent time in Bern and Madrid to discuss increasing the size of Swiss and Spanish use, and obtained an agreement with Italy, though at the time the existing partners could not agree that it be signed. The new democracies of Eastern Europe are waiting: I hope my successors at ILL can carry forward the discussion I started for an 'association' agreement with the Hungarian Academy of Science. ILL has long had exchange agreements with the Soviet Union which one day may ripen into full partnership. But the UK must be fully part of this policy making, closely engaged and pressing its views strongly. For the ILL is a crucible, and when you mix up the ingredients in such a vessel, the result is not just finely mixed ingredients, but something new. The dish is well worth preparing and savouring.

part 2

SOME PAST MASTERS

Heroic times breed heroic people. In science, as much as in politics or the arts, circumstances call forth the talents of those who are ready to seize the chance when it comes, and in the form that it takes. Each era, moreover, delivers its own quota of challenges and opportunities. For science (though the word was not current at the time—read 'natural philosophy'), the last decade of the eighteenth century and the first two of the nineteenth was a period of rapid change. In chemistry alone, the overthrow of phlogiston and the coming of the idea of chemical elements and what we now call the Periodic Table unlocked movements of intellectual tectonic plates that start to make the world of the physical sciences comprehensible to a 21st century observer, much as Chaucer's Canterbury Tales starts to give us an impression of contemporary English.

Because, leaving aside the Royal Society and its congenors among other national academies of the time, the Royal Institution was pretty much the only scientific forum around at the time, it was a focus for this ferment. Those who founded and guided it contributed uniquely to establishing an ethos for science that guided it up to the present time. The three great progenitors of the Royal Institution were Benjamin Thompson (better known as Count Rumford), Humphry Davy and Michael Faraday, each a towering genius in his own right but otherwise entirely disparate (indeed, most probably incompatible) in character and temperament. Each one has been the subject of numerous biographies and learned articles by professional historians of science, to the extent that it is hard to imagine any aspects of their lives and works that have not been subjected to detailed scrutiny and analysis. Nevertheless there remain corners of their careers that do not appear to have attracted so much attention. The object of the following chapters is to bring forward some lesser known facets of these almost mythic figures, putting some flesh on the scuptural bones and rendering them more real and human in a present-day context.

Before (and indeed after) he arrived at the Royal Institution, Rumford led a peripatetic and colourful life—'a girl in every port' would not be too exaggerated an epithet. Davy had many problems, personal as well as scientific, while Faraday, as the junior 'gopher', was put to work by his boss Davy on topics that were not especially congenial to him (though essential for bringing money into the Royal Institution) but which, nevertheless, he tackled with all the skill and insight that, later on, he was to bring to bear on more momentous matters.

chapter 4

Count Rumford's European Travels

There can be few scientists in history who have had the honour of being elevated to the status of Count of the Holy Roman Empire. Even fewer must be the holders of such an aristocratic European title who have been British citizens. And if it is possible, still fewer are likely to have been born in North America. Yet such an unlikely conjunction of circumstances occurred in the case of Benjamin Thompson, born in his grandfather's farmhouse in Woburn, Massachusetts, on 26 March 1753. He has been described by the science historian W. H. Brock as:

> *A loyalist, traitor, spy, cryptographer, opportunist, womaniser, philanthropist, egotistical bore, soldier of fortune, military and technical advisor, inventor, plagiarist, expert on heat (especially fireplaces and ovens) and founder of the world's greatest showplace for the popularisation of science, the Royal Institution.*

Not that my purpose here is to look into all (or indeed many) of this striking list of epithets. What I wish to do is to add yet another: 'distinguished European'. For Benjamin Thompson, better known as Count Rumford, was truly a citizen of Europe, at home and recognized in many countries, with friends and contacts everywhere and the instigator of projects in several capitals. A member of the Academies of Sciences of Bavaria (1786) and Berlin (1787), he had already been elected a Fellow of the Royal Society in 1778. Later,. in 1803, he was proposed by the 'Class of Mathematics and Physics' for election as one of the 24 foreign members of the Institut in Paris. Worldly honours came to him as well as academic ones. In 1783, he had been knighted by King George III of England; in 1786, the King of Poland conferred on him the Order of St. Stanislaus, while in 1788, his then employer, the Elector of Bavaria, appointed him Major General of Cavalry and Privy Counsellor of State. Finally, and most auspiciously, in 1791, when there was an interregnum between the death of the Emperor Joseph and the coronation of Leopold II, during which the Elector of Bavaria was one of the Vicars of the Empire, the latter took the opportunity of making Sir Benjamin Thompson a Count of the Holy Roman Empire, and at the same time invested him with the Order of the White Eagle. For what remarkable accomplishments were these multifarious accolades accorded, and by what means did the son of a farming family from rural Massachusetts come to be in a position to render such diverse services, while at the same time achieving a lasting reputation as an experimental physicist?

The guiding principle of Rumford's life, energy, single-mindedness, inquisitive curiosity and insensitivity to the opinions of others, appeared early in his career. He first went to school in Woburn where it is said that he neglected regular work but liked arithmetic. Leaving school at 13, he was apprenticed to an importer of British goods but 'instead of watching for customers over the counter, he busied himself with tools and instruments under it', an activity that included inventing a perpetual motion machine. In the period leading up to the War of Independence certain merchants, including the one to which the young Benjamin Thompson was apprenticed, signed an agreement not to trade with the United Kingdom. Thus, his apprenticeship was rendered otiose, but after learning French at evening school he continued his education (among other means) by attending lectures on Experimental Philosophy at Harvard College, while receiving personal tuition in anatomy, chemistry and medicine. It is at this point that the name by which he has subsequently became known appears in his life for the first time: he began to teach in a school in Concord, New Hampshire, which had earlier been called Rumford when it formed part of Essex County, Massachusetts. The name was changed to the one we know today after the ending of a dispute about county and state boundaries.

It was at Concord when, still not quite 20 years old, he met and married Mrs Rolfe, a rich widow ten years older than himself or, as he later told his friend Professor Pictet of Geneva, that she married him rather than he her. He was already a striking personality: one of his friends described him as 'of fine, manly make and figure, nearly six feet high, with handsome features, bright blue eyes and dark auburn hair. His manners were polished and his ways fascinating, and he could make himself agreeable. He had well used his opportunities of culture, so that his knowledge was beyond that of most of those around him'. He soon gave up school teaching and, as a consequence of his new wife's acquaintance with the State Governor, obtained a commission as Major in the New Hampshire regiment. The attainment of such a high rank by one so new to the military caused extreme resentment on the part of the regiment's junior officers and certainly helped to bring about the circumstances that eventually made it necessary for him to leave North America.

The War of Independence was about to begin. Among those who had worked for him on his family farm were four deserters from the British army in Boston. Thompson persuaded them to return to their regiment and used his acquaintance with the Governor to secure them a pardon. For this he was called before a committee of the people in Concord for being 'unfriendly to the cause of liberty'. A mob gathered at his house but he was able to get away unhurt. As the civil conflict developed, he temporized with the insurgents and apparently thought of gaining a commission from General Washington, but finally decided to leave for Boston. With the rebellion turning into revolution, Boston was evacuated shortly after by the British and Major Thompson was sent to England with the news.

Casting a glance forward to the extraordinary developments of his later life, it is worth noting that one of the sources for assessing his character and attributes at the various stages of his career is the '*eloge*' spoken by Baron Cuvier to the French Academy following his death in Paris in August 1814. Commenting on the Thompson of 1776, Cuvier said '*the good bearing of the young officer and the clarity and extent of the intelligence that he furnished told in his favour with the Secretary of State for the American Department*'. The latter was Lord George Germain, who appointed Major Thompson to no less a post than Secretary of the Province of Georgia. It was while staying with him at his country house that Thompson made his first observations on the heat generated in a gun barrel when it was fired with gunpowder. In short, he found that the exploding gunpowder made the outside of the barrel much hotter when there was no bullet to be fired than when the gun was loaded. This is probably the first direct observation that mechanical energy and heat are related, and can be converted one to the other. Only a few months later he was elected a Fellow of the Royal Society, and became a Lieutenant-Colonel.

The next 18 months of Colonel Thompson's life were passed in North America, where he was put in charge of a cavalry regiment fighting from North Carolina to Long Island, until (the British army being now defeated) those soldiers of the regiment who wished to remain in the country were allowed to keep their rank and were pensioned off on half pay, an advantage which Colonel Thompson kept for the rest of his life.

Nothing in what has been recounted so far, apart from Thompson's versatility and easy adaptation to new challenges, could be taken as foreshadowing the life that he was subsequently to make for himself in Europe. However, the abrupt end to his career in the British Army forced him to reconsider his future. With typical adventurousness, he decided to put his military training to good use by travelling to take part in the war that was expected to break out between Austria and the Ottoman Empire. It was at this juncture that the British historian Edward Gibbon was able to write from Dover in 1783:

> *Last night the wind was so high that the vessel could not stir from the harbour; this day it is brisk and fair. We are flattered with the hope of making Calais Harbour by the same tide in three hours and a half, but any delay will leave the disagreeable option of a tottering boat or a tossing night. What a cursed thing to live in an island: This step is more awkward than the whole journey. The triumvirate of this memorable embarkation will consist of the grand Gibbon, Henry Laurens, Esq., President of Congress, and Mr Secretary, Colonel, Admiral, Philosopher Thompson, attended by three horses, who are not the most agreeable fellow-passengers. If we survive, I will finish and seal my letter at Calais. Our salvation shall be ascribed to the prayers of my lady and aunt, for I do believe they both pray.*

As Thompson travelled across Europe, a purely accidental meeting was to change the course of his life. The circumstances were described by Pictet in the 'Bibliotheque Universelle':

> *A purely accidental circumstance had a decisive influence over his destiny. He arrived at Strasburg, where the Prince Maximilian of Deux Ponts, now [1801] Elector of Bavaria, then Field Marshal in the service of France, was in garrison. This prince, commanding on parade, sees among the spectators an officer in a foreign uniform, mounted on a fine English horse, whom he addresses. Thompson informs him that he comes from serving in the American war: The Prince, in pointing out to him many officers who surround him, says, 'These gentlemen were in the same way but against you; they belonged to the Royal Regiment of Deux Ponts, that acted in America under the orders of Count Rochambeau...'. When at last the traveller took leave, the Prince engaged him to pass through Munich, and gave him a friendly letter to the Elector of Bavaria, his uncle.*

In February 1784, Thompson was knighted by King George III and given permission to enter the service of the Elector of Bavaria: he described himself as 'M. le Chevalier Thompson, colonel and general aide-de-camp in the service of His Imperial Highness the Elector Palatine, Duke of Bavaria'. His European odessey had begun.

The Bavaria into whose service M. le Chevalier Thompson entered was, from the viewpoint of one whose independent thoughts had been honed on the civil and administrative systems of England and its erstwhile North American colonies, a pretty backward place. To quote Cuvier's eloge again:

> *Because of their devotion to the catholic faith, the rulers who flourished at the time of the religious wars had long carried the imprint of their fervour well beyond what was required by an enlightened Catholicism; they encouraged piety but did nothing for manufacture; throughout their kingdoms could be found more convents than factories. The Army was more or less non-existent; ignorance and apathy dominated all levels of society.*

Straightaway Thompson set about reorganising the Elector's army. His view about the conjunction of civil and military life is summarised in the following statement:

> *I was ever mindful of that great and important truth that no political arrangement can be really good except in so far as it contributes to the general good of society. I have endeavoured to unite the interest of the soldier with the interest of civil society, and to render the military force, even in the times of peace, subservient to the public good.*

He used the army as a means of improving both agriculture and manufacture; gardens for growing potatoes and workshops for producing uniforms were introduced, first in Mannheim and then in Munich. The enterprise even ran at a profit. Further challenges followed. The country was filled with indigent poor, whose distress could be relieved by using the peacetime resources of the army, not just by clearing them off the streets but providing work for the able and help to those unable to fend for themselves. Thompson's positive approach to social problems (acquired from his American upbringing?) is beautifully encapsulated in his prescription: 'to make vicious and abandoned people happy, it has generally been supposed necessary first to make them virtuous. But why not reverse this order? Why not make them first happy and then virtuous?' A poorhouse was organised, with kitchens and workshops so that the poor could carry out their own cookery and earn modest amounts of money from their work. One objective of the latter was to provide clothing for the army. Thompson adopted a combination of carrots and sticks to encourage hard work and good performance:

> *To incite activity and inspire with a true spirit of persevering industry, it was necessary to fire the poor with emulation—to awaken in them a dormant passion whose influence they had never felt; the love of honest fame; an ardent desire to excel, the love of glory, or by what other pompous name this passion, the most noble and most beneficent that warms the human heart, can be distinguished.*

Thompson's efforts were so greatly appreciated that when he became ill, he was the subject of intense prayers; as told in his own words:

> *Let the reader, if he can, picture my situation. Sick in bed, worn out by intense application, and dying, as everybody thought, a martyr in the cause to which I had devoted myself, let him imagine, I say, my feelings upon hearing the confused noise of the prayers of a multitude of people, who were passing by in the streets, upon being told that it was the poor of Munich, many hundreds in number, who were going in procession to the church to put up public prayers for me; for a private person, a stranger, a Protestant!*

Several other projects for the state of Bavaria lend further testimony to Thompson's organising abilities. He oversaw the establishment of a Military Academy, to which the offspring even of the poorest of the Elector's subjects could be educated, provided they showed evidence of strong aptitude. He employed army labour to improve the roads, and even set about improving the breeding of horses and cattle. One lasting monument to his endeavours is the English Garden in Munich, laid out on the old town ramparts. Within it was originally housed a model farm, and in the centre a coffee house. It was in recognition of all these manifest improvements to his realm that the Elector of Bavaria gave Sir Benjamin Thompson the name and title by which he has since been best known: Count Rumford.

Rumford's many projects were not only carried forward with quite single-minded zeal, but with careful planning and, perhaps most important of all, minute attention to the nature of the materials and techniques being employed. The latter led him, through careful observation and experiment, to the advances in physics with which his name is still associated. As Cuvier said:

> *In fact it was through working for the poor that he made his greatest discoveries... We all know that his first experiments were directed towards the nature of heat and light and the laws governing their propagation, and it was through that endeavour that he arrived at a better understanding of how to feed, clothe, warm and light a large body of people economically.*

Feeding and heating, in particular, provide two remarkable examples of Rumford's diligent pursuit of the application of quantitative logical analysis to practical problems. Until Rumford analysed the problem, the poor people who received their nourishment from the soup provided at the House of Industry in Munich were a considerable charge on the Bavarian state. By noting quite precisely how much of each ingredient was necessary, and in particular by introducing potatoes as a substitute for the more expensive pearl barley, the cost per portion was reduced to one farthing, including wages and fuel! Furthermore, after initial suspicion, the consumers apparently thought their fate had been greatly improved by the innovation. Rumford described the circumstances:

> *But, moderate as these expenses are which have attended the feeding of the poor of Munich, they have lately been reduced still farther by introducing the use of potatoes. These most valuable vegetables were hardly known in Bavaria till very lately; and so strong was the aversion of the public, and particularly of the poor, against them, at the time when we began to make use of them in the public kitchen of the House of Industry in Munich, that we were absolutely obliged, at first, to introduce them by stealth. A private room in a retired corner was fitted up as a kitchen for cooking them; and it was necessary to disguise them by boiling them down entirely, and destroying their form and texture, to prevent their being detected. But the poor soon found that their soup was improved in its qualities; and they testified their approbation of the change that had been made in it so generally and loudly that it was at last thought to be no longer necessary to conceal from them the secret of its composition, and they are now grown so fond of potatoes that they would not easily be satisfied without them.*

Other additions, of meat and bread, were made for flavour rather than nourishment, in a highly calculated way:

> *As the meat in these compositions is designed rather to please the palate than for anything else, the soup being sufficiently nourishing without it, it is of much importance that it be reduced to very small*

pieces, in order that it be brought into contact with the organs of taste by a large surface; and that it be mixed with some hard substance (fried bread, for instance, crumbs, or hard dumplings), which will necessarily prolong the time employed in mastication... When this is done, and where the meat employed has much flavour, a very small quantity of it will be found sufficient to answer the purpose required.

Rumford's contribution to the science and applications of heat arose, on the one hand, from his work with the poor and, on the other, from his military connections. To economise still further on the cost of looking after the poor, he invented a new kind of stove, the forerunner of the convector heater. Around the wood or coal-burning compartment, which had an open front of a conventional kind, was fixed a further enclosure with apertures below and above so that cool air drawn in at the bottom would be heated as it rose by convection around the stove and expelled into the room from the top. A contemporary cartoon of Rumford illustrates how famous this invention became (Fig. 27).

But the scientific discovery for which he remains most famous was made while he was superintending the boring of cannon barrels in the foundry he had established in Munich. He was struck by the way in which the barrel itself became hot, and still more by the intense heat of the metal shavings. As he put it:

The more I meditated on these phenomena, the more they appeared to me to be curious and interesting. A thorough investigation of them seemed even to bid fair to give a further insight into the hidden nature of heat, and to enable us to form some reasonable conjectures respecting the existence or non-existence of an igneous fluid.

His conclusion gives us the first glimpse of a modern theory of heat:

Anything which any insulated body or system of bodies can continue to furnish without limitation cannot possibly be a material substance, and it appears to me to be extremely difficult, if not quite impossible, to form any distinct idea of anything capable of

Fig. 27. The comforts of a Rumford stove (a cartoon by Gillray).

being excited and communicated in these experiments except it be MOTION... I am far pretending to know how that particular kind of motion which has been supposed to constitute heat is excited, continued, and propagated. Nobody surely in his sober senses has ever pretended to understand the mechanism of gravitation, and yet what sublime discovery was our immortal Newton enabled to make merely by the investigation of the laws of its action!

Rumford published a series of articles about his scientific and administrative work, the former in the Philosophical Transactions of the Royal Society of London and the latter in a set of volumes entitled 'Essays, Political, Economical and Philosophical'. It is also significant that some of them appeared separately in French in '*Receuil de Memoires sur les Etablissements d'Humanite*' published by Henri A. Gasse in An VII of the revolutionary calendar. One of these articles contains proposals for founding an establishment in London and so contains the germ of the idea that was later to be manifested as the Royal Institution.

It is not my purpose here to retell the story of how the Royal Institution was founded, since that has been done earlier in this book. What is interesting in the present context, though, is the way that the concept of this unique organisation, which seeks to combine the discovery of new scientific knowledge about matters impinging on everyday life with its dissemination to a wider audience, arose out of Rumford's diligent work in Bavaria. Of course, the character of the Royal Institution has evolved greatly since 1799, though it is still housed in the premises bought for the purpose in Albemarle Street, close to Piccadilly, in that year. Nevertheless, the 'mission statement' contained in the Royal Charter by which it was formally established (and which clearly shows evidence of Rumford's drafting) still encapsulates its unique combination of functions:

> *... for diffusing the knowledge and facilitating the introduction of useful mechanical inventions and improvements, and for teaching by courses of philosophical lectures, and experiments, the application of science to the common purposes of life.*

In writing the prospectus, assembling and organising the group of subscribers, purchasing the house, hiring staff (including architects and workmen), Rumford's prodigious energy was fully stretched in the years on either side of 1800 as the Royal Institution came into being in the fashionable heart of Mayfair. Yet such a task could only have been accomplished so thoroughly by a person of distinctly autocratic temperament. Rumford had numerous disagreements with the Committee of Managers, and caused great distress and difficulty to the first lecturer at the new

Institution, Dr Thomas Garnett. A contemporary satirical poet, writing under the pseudonym Peter Pindar, had this to say of him:

But what in insolence in me to prate,
Pretend to him to open Wisdom's gate,
Who spurns advice, like weeds, where'er it springs,
Disdaining counsel, though it comes from kings.

It has even been said that the root of the next phase in his European odyssey, this time to Paris, was the continuing argument about the future and policy of the Royal Institution. However, this appears not to be so: rather, a commission from the Elector of Bavaria gave him immediate cause to revisit Munich, but most likely the real influence was that of Mme Lavoisier, the widow of the celebrated chemist Antoine Lavoisier, guillotined in 1794. A thumbnail sketch of this remarkable lady was drawn by M. Guizot in 1841, five years after her death:

Fondness for her husband, as well as her own innate inclinations, led Mrs Lavoisier to take part in his work as a comrade and a disciple. Those who only knew her when she was no longer a young woman could be forgiven for not noticing that under a somewhat cool and formidable exterior, almost entirely taken up with social matters, was a personality capable of being strongly moved by feelings and idea, and give herself over to them passionately. A private life lit up equally by reciprocal affection and favourite pastimes, a massive fortune, high esteem, a beautiful house at l'Arsenal, sought after by people of the highest distinction, all the pleasures of intellect, riches and youth—it was without doubt an enviable and easy life. That life was fractured, indeed torn apart by the revolution, like that of all those around her. On the same day in 1794, Mrs Lavoisier saw both her father and husband mount the scaffold and she herself only escaped, after a brief period in prison, by hiding herself in complete and silent obscurity.

When these constraints came to an end, and order and justice returned to revive and pacify society, Mrs Lavoisier took her place in the world again, surrounded by a whole generation of famous intellectuals who had been friends, disciples and successors of Lavoisier. Lagrange, Laplace, Berthollet, Cuvier, Prony, Humboldt and Agago, all delighted in honouring Lavoisier's widow, finding in her house a return to the brilliance that they remembered, combined with the pleasures of elegant hospitality. Then Rumford arrived among them. He found favour with Mrs Lavoisier; he was in tune with her tastes and habits—one might almost say, with her memories.

The first time Rumford met Mme Lavoisier was in Paris, in October 1801, during a visit when he was also introduced to Napoleon himself, then First Consul, at a meeting of the Institut, to which Volta gave a presentation of his discovery of the voltaic battery. Rumford described the occasion, and his assessment of Bonaparte, in a letter to Sir Joseph Banks:

After Volta had finished his presentation the First Consul demanded leave from the President to speak, which, being granted, he proposed to the meeting to reward M. Volta with a gold medal, and to appoint a committee to confer with M. Volta on the subject of his experiments and investigations respecting galvanism, and to make such new experiments as may bid fair to lead to further discoveries. He delivered his sentiments with great perspicuity and displayed a degree of eloquence which surprised me. He is certainly a very extraordinary man and is possessed of uncommon abilities. The expression of his countenance is strong, and it is easy to perceive by his looks that he can pronounce the magic words 'je le veux' with due energy.

During the same visit, he made the acquaintance of leading scientists, including Laplace and Berthollet; his fame as an administrator also reached the higher levels of French political life, and he was invited to dine both with Chaptal, the Minister of the Interior, and the Minister for Foreign Affairs, Talleyrand.

Moving between the two parts of the Elector's realms, in Munich and Mannheim, alternating with extending visits to Paris, Rumford was more than once joined by Mme Lavoisier. As Sir C. Blagden wrote from Paris to Sir Joseph Banks: 'Count Rumford arrived in remarkably good health ... travelling agrees with him'. Although the same correspondent, writing this time to Rumford's daughter, stated in August 1803 that 'I am still as much at a loss as I was in June to answer your question whether your father be going to marry'. By January 1804 the matter was clear. Rumford himself wrote to his daughter:

> *I shall withhold this information from you no longer. I really do think of marrying, though I am not yet absolutely determined on matrimony. I made the acquaintance of this very amiable woman in Paris, who, I believe, would have no objection in having me for a husband, and who in all respects would be a proper match for me. She is a widow, without children, never having had any, is about my own age, enjoys good health, is very pleasant in society, has a handsome fortune at her own disposal, enjoys a most respectable reputation, keeps a good house, which is frequented by all the first philosophers and men of eminence in the science and literature of the age, or rather of Paris, and, what is more than all the rest, is goodness itself... She has been very handsome in her day, and even now, at forty-six or forty-eight, is not bad-looking; of a middling size, but rather 'en bon point' than thin. She has a great deal of vivacity and writes incomparably well.*

Some financial arrangements between the couple followed soon after, but it was not until the following October that the marriage finally took place. However, through his description to his daughter of Mme Lavoisier's house (where they set up home together), clear signs of the egotism that was to strain the marriage shine through. From 39 rue d'Anjou, Paris, he wrote:

> *I have the best-founded hopes of passing my days in peace and quiet in this paradise of a place, made what it is by me—my money, skill, and directions.*

An early source of contention between them arose quite simply from the name by which the former Mme Lavoisier wished to be known. In the marriage contract it appears that she stipulated quite formally that she should be called Mme Lavoisier de Rumford. In her own words:

> *I have thought it an obligation, almost a religion, never to give up the name of Lavoisier. Counting on the word of Mr Rumford, I would*

never have made it a clause of my civil contract with him had I not wished to make a public statement of my respect for Mr Lavoisier and a proof of Mr Rumford's generosity. I consider it a duty to hold determinedly to what has always been one of the conditions of our union; and in the depth of my soul I have the inmost conviction that Mr Rumford will not disrespect me for this and that, having taken the time to think it over, he will allow me to continue fulfilling a duty I regard as sacred.

Rumford's letters to his daughter to America give us some graphic insight into his disagreements with his wife. The root cause of the disaffection between them seems quite simply to have been their independent frames of mind: each had lived alone for too many years to find the give and take of domestic family life at all congenial. 'Little it matters with me,' he writes, 'but I call her a female dragon—simply by that gentle name! We have got to the pitch of my insisting on one thing, she on another'. And again, 'I have the misfortune to be married to one of the most imperious, tyrannical, unfeeling women that ever existed, and whose perseverance in pursuing an object is equal to her profound cunning and wickedness in framing it'. In these confrontations comedy was sometimes not far away:

A large party had been invited I neither liked nor approved of, and invited for the sole purpose of vexing me. Our house being in the centre of the garden, walled around, with iron gates, I put on my hat, walked down to the porter's lodge and gave him orders, on his peril, not to let anyone in. Besides, I took away the keys. Madame went down, and when the company arrived she talked with them, she on one side, they on the other, of the high brick wall. After that she goes and pours boiling water on some of my beautiful flowers.

Separation was the only solution: it took place officially on 30 June 1809, after which Rumford became, as he put it, his own man again. He bought the lease of a villa at Auteuil, between the Seine and the Bois de Boulogne. He continued to see his former wife from time to time; indeed one has the distinct impression that they became much better reconciled to one another when no longer living under the same roof! Rumford's daughter visited him, remarking that the great difference between he and his ex-wife was that 'he was fond of experiments, and she of company'. Experiments, indeed, continued to occupy him, and he read a paper before the Institut on 'Heat Manifested in the Combustion of Inflammable Substances'. During his continental tour of 1812–1814, so beautifully described in the journal kept by the young assistant, Michael Faraday, Humphry Davy dined with Rumford at Auteuil. No doubt Davy was able to recall how, 12 years previously, he had been hired by the founding father of the Royal Institution as Assistant Lecturer there, the beginning of his fame in science.

Rumford died at Auteuil on 25 August 1814, having never returned to Britain since 1803. At the beginning of 1815, it fell to Baron Cuvier to read his eulogy to the Academie Française, in which some of the sharper edges of his character are forcefully evoked, while praising his manifest achievements:

> *In fact the last ten years saw him honoured by the French and by foreigners, admired by friends of science, sharing in their work, helping even the humblest craftsman and amply rewarding the public with all the useful things he invented day by day. Nothing would have been lacking from the pleasure of his life if the smoothness of his social demeanour had equaled his enthusiasm for the public good. But we must admit that his conversation and manner left an impression quite extraordinary in one who had always been well treated by others and who himself had done them so much good. It must be said that it was in fact without either liking them or valuing them that he rendered his services to his fellows.*

This unusual, angular, driven and ultimately sad personality left behind the esteem of both scientists and politicians in Britain, France and Bavaria. Apart from his science, tangible memorials to him remain to this day in the English Garden at Munich, the naming of one of the premier medals of the Royal Society and above all, the Royal Institution in London. Cuvier's final words sum up this remarkable European:

> *A man who, by a felicitous choice of topics for his work, succeeded at the same time in attracting the high regard of intellectuals and the gratitude of the poor.*

Note: The translations from the French are by the author.

chapter 5

Humphry Davy's Quest for Research Funding

Love of money may be the root of all evil, but it remains a topic of perennial interest to scientists. Except for those theorists so arcane that they need only pencil and paper, finding the wherewithal to do the next experiment has always been a major preoccupation for practising scientists, and the sources to be tapped are correspondingly disparate. Nowadays we think of governments and industry as the main providers of research money, but in earlier times it was private generosity that scientists usually appealed to. Many scientists were men of aristocratic lineage, which gave them both the personal resources and the leisure to pursue intellectual pastimes. Intellectual is the appropriate word, because projects likely to lead to immediate profit came under the heading of 'arts and manufactures'. For example, the third Lord Rayleigh's work on separating the rare gases was done in his own laboratory at his country house in Essex—which is now being renovated and catalogued. Today we take it for granted that the public purse should support research that brings no immediate pay-off: 'curiosity driven' or 'basic' are the adjectives in vogue. But how did such a view come about? The vehicle for requesting money for a research project is nowadays the grant proposal. A scientist fills in a form detailing the various expenses to be incurred under the headings of equipment, consumables and salaries, together with information about the timescale and likely outcome, and sends it off to one of the Research Councils or research charities.

Fig. 28. Humphry Davy (portrait by Thomas Lawrence at the Royal Institution).

For those who have to decide whether to grant funds for the proposal, the most important part of the documentation is the 'case for support'. In this context, what is arguably the first ever grant proposal for a piece of basic research opens a revealing window on the way that some of the greatest science in British history—Humphry Davy's (Fig. 28) electrochemical research—was justified and financed. Echoes of our present strategies for balancing budgets are easy to see. The proposal in question was made to the Managers of the Royal Institution, one of the first public organisations to be set up anywhere in the world with the aim of enlarging knowledge through experiments. In fact, its earliest goals were quite utilitarian: 'the application of science to the common purposes of life' is the famous phrase in the 1800 Charter. A report to the Managers of the Royal Institution at the end of April 1803 gives a vivid account of the early state of affairs as far as experimental sciences was concerned:

> *Chemistry was always a primary objective of the Institution. A laboratory was therefore erected at an early period and was furnished with such equipment as was immediately requisite; but, as the Institution was in a nascent state, great attention was paid to economy in the chemical department. On this account the laboratory has remained in a state inferior to that which might justly be expected in such a liberal and splendid establishment.*

Accordingly, detailed proposals were put forward to remedy this situation by enlarging and equipping the laboratory so that, in the words of the report, 'it shall be equal, or indeed superior, to any in this country, and probably in any on the Continent'. Action was immediate: on 2 May it was resolved that a 'Committee of Science' should be appointed from among the Managers, and on 16 May, the Managers resolved 'that the Committee of Science should carry out improvements in the laboratory'.

For the next decade, the Royal Institution's laboratory was the scene of some of the most important work in the history of chemistry, namely Davy's experiments on electrochemistry, culminating in his isolating sodium, potassium, magnesium, calcium, strontium and barium. These not only brought great fame to the laboratory, but provided the impetus for what is perhaps the earliest formal research funding proposal in the history of science. However, before telling that story it is worth noting other ways in which the experimental research programme was conducted and supported. Principal among what we would now call 'externally funded programmes' was chemical analysis, especially involving topics connected with agriculture. The President of the Royal Society, Sir Joseph Banks, was closely involved in setting up the Royal Institution, and his position gave him access to government. On 27 May 1803, he wrote on behalf of the Committee of Science to the Board of Agriculture:

> *The Committee are aware that at present the science of agricultural chemistry is in its infancy and that until it has been more matured each analysis will take up considerable portion of time; they trust however that it will not be long before Mr Davy himself, or some one named by him and acting under his superintendence, will undertake the business of analysing soils and manures for individuals at a moderate fixed price for each substance that shall be brought to them.*

It is clear that, unlike some who propose providing a service in return for a fee, Banks was at pains not to appear to be offering more than he thought the Royal Institution could deliver, because he adds as a rider:

> *The Committee do not expect in agricultural analysis the same degree of precise accuracy as is necessary in that intended to illustrate philosophical experiments; it will be enough for them if the component parts of substances and their respective proportions to each other are marked with sufficient precision to demonstrate the probable effects on vegetables.*

The word 'philosophical' denotes what we would now call basic science, i.e. work undertaken for its own intrinsic interest, to uncover the laws of nature.

Banks's proposal for funding from the Board of Agriculture contained two elements, both representing contributions to Davy's salary. The first was a one-off payment of 60 guineas for six lectures, which Davy delivered at the Board; the second an on-going obligation to underwrite Davy's earnings at the rate of £100 a year, in return for which he would have to give lectures each spring on 'the application of chemistry to the improvement of the art of agriculture', and also undertake the analytical work. For this he was to be given the title of 'Professor of Chemical Agriculture to the Board'. No doubt these arrangements were what Bernard meant when, in a report in May 1803, he wrote that 'the new plan for the laboratory promises to increase the scope and vitality of it and at the same time very much to diminish, if not eventually provide for, the expense of that part of the Institution'.

In the following year, access to the laboratory was opened up to a wider public, to anybody who paid the fee of £10. So contract research, sponsored lectures and partial underwriting of personnel costs from an outside agency contributed to the financial rescue of this scientific organisation. Added to that, the Members, Managers, Secretary and Treasurer also stumped up the not inconsiderable sum of £100 each, as an interest-free loan.

With the financial breathing space gained by all these manoeuvres, the Royal Institution continued to survive, and even flourish. Davy gave his lectures on agricultural chemistry (they were finally published as a book in 1813), but his own personal research was burgeoning in quite another direction: electrochemistry or, as it was then called, galvanism. As early as October 1800, when he was still at the Pneumatic Institution in Clifton, Bristol, Davy wrote to one of his friends, 'galvanism I have found, by numerous experiments, to be a process purely chemical'. As soon as he arrived at the Royal Institution in March 1801, he set out to continue and augment the work on electrochemistry that he had started in Clifton.

That is where we meet another characteristic of scientific progress that has plagued research funding down the ages—the sophistication factor. Starting from single metal plates, following the principle developed by Volta, Davy was first using very simple equipment. For example, in November 1801 he found he could make a battery without using metal electrodes. As he put it 'by means of ten pieces of well-burnt charcoal, nitrous oxide and water, arranged alternately in wine glasses, I produced all the effects usually obtained from zinc, silver and water'. However, by the moment of 1802, we find him writing that he had 'lately had constructed for the laboratory of the Institution a battery of immense size; it consists of four hundred plates of five inches in diameter and forty of a foot in diameter'. The vast equipment illustrated in Fig. 29 may represent this battery, installed in the basement of the Royal Institution, or it may be an artist's impression of Davy's much greater project for the future.

DAVY'S FIRST GRANT PROPOSAL

A new path of discovery having been opened in the agencies of the Electrical Battery of Volta, which promises to lead to the greatest improvements in Chemistry and Natural Philosophy, and the useful

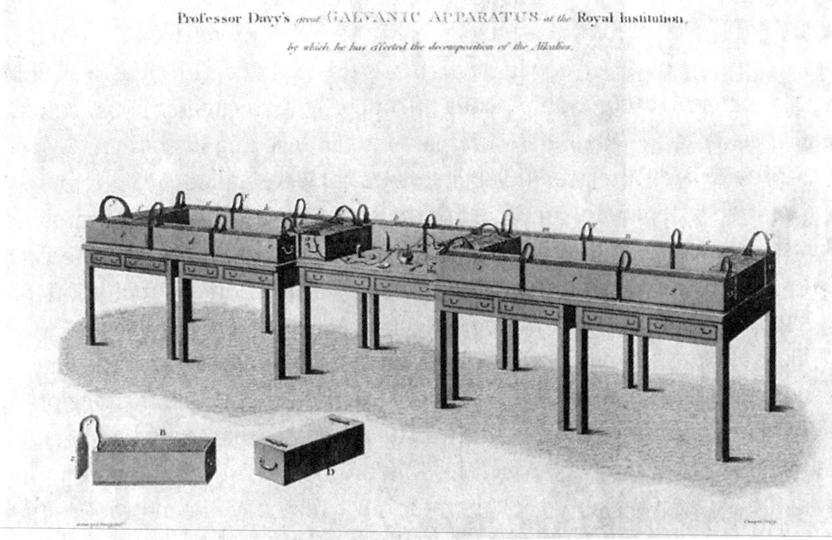

Fig. 29. Davy's battery, or an artist's impression of a future one.

Arts connected with them; and the increase of the size of the Apparatus being necessary for pursuing it to its full extent, it is proposed to raise a fund by subscription, for constructing a powerful Battery, worthy of a National Establishment, and capable of promoting the great objects of Science. Already in other Countries, public and ample means have been provided for pursuing these investigations. They have had their origin in this Country, and it would be dishonourable to a nation so great, so powerful, and so rich, if, from the want of pecuniary resources, they should be completed abroad. An appeal to enlightened individuals on this subject can scarcely be made in vain. It is proposed that the Instrument and apparatus be erected in the laboratory of the Royal Institution where it shall be employed in the advancement of this new department of Science.

11 July 1808

Because the lectures on agricultural chemistry and analytical work took up most of his time, Davy was not in a position to build on his earlier successes in electrochemistry for several years, but when he finally returned to his favourite topic, dramatic progress came at once: he decomposed water to yield hydrogen and oxygen in the proportions we know today. That experiment, described in detail in his first Bakerian Lecture, was done with a battery of 100 double plates of copper and zinc, each 15 cm square, and in agate rather than glass vessels to avoid leaching of sodium. All this was provided by the wealthy backers, who were called 'proprietors' or 'subscribers', depending on whether they paid life or annual subscriptions. Potassium, too, was first isolated within a month, and to crown the year, Davy was awarded the prize for the best experiment on electricity, offered by the Institut de France on the instruction of Napoleon.

Early in the following year, the Managers' minutes contain a sombre phrase: 'there appears to be an excess of expenditure beyond the receipts of income'. One contributing reason was undoubtedly that Davy had been ill, and hence unable to give the lectures that brought in subscriptions. Another was 'the extra expense of the laboratory in which have been produced Mr Davy's recent discoveries, so honourable to the Royal Institution and so beneficial to the Interests of Science in every part of the world'. This expense amounted to £166 10s—a large sum in those days.

Fortunately, Davy recovered from his illness and began to lecture again in the spring of 1808, a period that also brought another spate of discoveries: by the end of June he had isolated calcium, strontium, barium and magnesium. In addition to the appointment he already held as Professor of Chemistry, he was appointed 'Superintendent of the House' (what we would now call Director), and it may be that it was this new status that emboldened him to make a formal proposal to the Managers for research funding.

In the light of all that we now know about the art of writing research grant proposals, it is fascinating to dissect the strands of the argument that Davy put forward. The proposal is for an equipment grant 'for constructing a powerful battery', and it is justified in ways that will be entirely familiar to anyone who has made grant applications to Research Councils. He begins by drawing attention to the important discoveries that have already been made using batteries (perhaps what would be called nowadays 'feasibility studies') and mentions that they have not only advanced basic understanding ('natural philosophy') but—and there is a very modem resonance here—also hold promise of 'useful arts'—applications! He then moves on to invoke the sophistication factor, and indicate that such advanced equipment should be situated in a 'national establishment' (for the latter, read 'centre of excellence'?), and perhaps there is even a hint that it might be a central facility.

Another familiar theme follows: nationalism, and the spectre of being left behind. The initial experiments giving rise to the proposal 'have had their origins in this country', he says—perhaps not entirely justifiably, in view of the fact that the battery bears the name of an Italian, Volta. But never mind; who could say that hyperbole is never found in grant applications? He then says that it would be 'dishonourable' if a great nation should be sidelined in the new field by lack of investment (somehow, 'want of pecuniary resources' sounds better). Finally, he appeals to the good sense of his interlocutors, whose reputation for sound judgment could only be enhanced by contributing to such a worthwhile project.

The Managers were immediately convinced. Those present at the meeting agreed to subscribe, and ordered 'that a book be opened at the Steward's Office for the purpose of entering the names of all those who may wish to contribute to this important national object'. How much did they raise? A few weeks later the Managers ordered that the 'subscription books for prosecuting Mr Davy's discoveries' were to be put on the table in the public rooms and written up every evening. It has been said that the total subscribed was £420. In any event, nine months later £86 13s 6d was paid to Mr Charles Royce for apparatus, of which £34 was specifically for 'the Voltaic battery' and shortly after, amounts totalling £286 9s 9d were paid to three other suppliers for 'galvanic apparatus'.

Davy got his battery, and the seeds of 'big science' were sown. Davy thanked the donors in a public lecture, which shows clearly how he thought basic science should be regarded, as a fitting object for disinterested support:

> *A scientific institution ought no more to be made an object of profit than an hospital, or a charitable establishment. What this Institution has done ... has tended to the progress of knowledge and invention ... with more ample support, more, undoubtedly would be affected ... it is no mean object to the country, that the first attempt of this kind should succeed.*

The history of science shows that it did succeed, not only through the new science it generated but through the seeds of a new attitude towards the public funding of basic research that Davy's 'first attempt' produced. As researchers in UK universities sit down to write their grant proposals to the Research Councils, they can feel that they are following the pathway (and perhaps even some of the lines of justification) that Humphry Davy mapped out in 1808.

chapter 6

Michael Faraday as a Materials Scientist

Michael Faraday was the greatest experimental scientist who ever lived. That bold claim is substantiated not just by the laws of electrolysis and magneto-optical rotation that bear his name, or the discovery of benzene and diamagnetism, or even, above all, electromagnetic induction. It is the sheer range, as well as the seminal significance of these discoveries that sets Faraday's research career on a pinnacle higher than any other. Yet the list is not complete. The bookbinder's apprentice who started his life in science as the chemical assistant to Humphry Davy, turning increasingly towards physics and finally laying the foundations of what we now know as field theory, spent many years establishing principles of yet another research domain, to which his masterly contribution is all but forgotten. The field in question is materials science.

Faraday's contribution to materials science has never been given much prominence by the historians who have documented and assessed his career: indeed, one of the most eminent of his biographers relegates all Faraday's work in the subject to a chapter entitled The Fallow Years. Yet it is also possible to argue that the skills that Faraday acquired during the years he worked to optimise the properties of materials, not to mention the insights he gained into the dependence of physical properties on composition (and especially trace impurities) gave him a grounding in observation and controlled experimentation that provide an essential basis for the larger glories that were to come. Of course, as the son of a blacksmith, the young Faraday was well placed to appreciate the physical properties of materials, and the way in which preparation and working could modify them. At the end of the twentieth century, when we can see and manipulate individual atoms with atomic force microscopes, and observe cracks propagating with electron microscopy, it is important to remember that all experiments in Faraday's time were concerned with the behaviour of matter in the bulk.

The objects that Faraday worked on all through his life were macroscopic—bar magnets, pieces of glass and metal—and the effects he saw did not need magnification. Nevertheless, it is certainly significant that he always looked for explanations for what he observed in changes in crystal form or structure. Is it here that we see his natural affinity with the discipline of materials science, concerned too with macroscopic behaviour (transparency, strength, . . .), but explicable only at the microscopic level? I want to illustrate Faraday's contribution to materials science through his work on two of the classical problems that have occupied materials scientists for generations, the hardening of steel and the formulation of flaw-free glass. Both show Faraday's supreme skill at observation, at analysing the causes of what he saw, and a marvellous ability to design simple but penetrating experiments to confirm his ideas.

It is certainly a truism that many of the most challenging problems in materials science arise from quite practical issues. That was the case in Faraday's time, too, and the projects on materials that he took up were brought to the Royal Institution as commissions, or, as we would say now, research contracts. The first arose in 1818 when James Stodart, a member of the Royal Institution, asked for help with analysing, and if possible reproducing, a hardened steel called wootz, imported from India. Samples of wootz had first been sent from India to Sir Joseph Banks, then President of the Royal Society, in 1795 but it had never been analysed. Mr Stodart kept a shop in the Strand selling cutlery, and especially razors, made from wootz so he had a clear interest in discovering its composition so that it could be manufactured in Britain (shades of import substitution?). Stodart had already experimented with steel himself, and had shown that the colours found on the surface of heated steel were the result of superficial oxidation, in contradiction to what Davy had believed. This he did by demonstrating that when steel was heated under an inert atmosphere, no colours resulted.

Under its second Director, Humphry Davy, the Royal Institution had become the leading British (and perhaps European) centre of chemical research, and given the remit handed to it by its founder Count Rumford to pursue 'the application of science to the common purposes of life', it was a natural choice for Mr Stodart's enquiry. By 1818 Davy was at the height of his fame and powers, having in turn become President of The Royal Society. Analysing steel samples failed to catch his interest, and the other Professor at the Royal Institution, Thomas Brande, was too busy preparing and delivering lectures. Thus the task was given to the young chemical assistant, Michael Faraday, whose early experience as a blacksmith's son must have made him appear a very suitable choice.

Faraday already had a solid reputation as an analytical chemist and he was a highly skilled experimentalist, so it would be marvellous to be able to report that he got to the heart of the problem at once: sadly, that is not the case. His first attempts at chemical analysis appeared to suggest that wootz contained large amounts of silicon and aluminium. In fact it does not, so when he tried to prepare it by adding these two elements to steel, the products bore no resemblance to wootz at all. To extend his knowledge about the practicalities of metal production he visited an iron foundry and, on a walking tour of Wales, a copper works. The latter was not a random choice because, from a report published in France of attempts to follow the path he had taken, Faraday concluded that it might be worth trying to harden steel in the same way as copper, namely by adding traces of noble metals (silver, gold, platinum). For this purpose, he needed a furnace capable of reaching exceptionally high temperatures and, true to his experimental tradition, he designed it himself.

The crucibles each consisted of two pots made from clay and graphite, one inside the other with the intervening space filled with a clay mixture. They were heated from below by a grate with provision for an air blast; pure iron could be melted in 12–15 minutes. After adding the different trace metals, Faraday checked whether the resulting ingots were homogeneous by etching them with acid, and it was here that he achieved one of his remarkably prescient insights that marked a turning point in the history of metallurgy. He concluded that the patterns formed on the surface by etching gave a picture of the underlying crystallites. Furthermore, he found that the damask surface so characteristic of hardened steel (indeed, the word comes from Damascus sword blades) was the result of crystals twisted and intertwined with one another by stress, for example by hammering.

The purpose of Faraday's experiments was not simply the pursuit of knowledge, but of commerce and, having successfully demonstrated that steel could be hardened by adding noble metals, experiments were worked up to pilot plant scale by transferring them to the foundry owned by Green, Pickslay and Co. in Sheffield. There (as letters from Pickslay to Faraday held at the Royal Institution testify) superintendence of the work was entrusted to an 'intelligent and confidential agent'. It would have made a splendid conclusion to this project in applied science to record that the British specialised steel industry grew mighty on the results of Faraday's research, but unfortunately it was not put into full scale production. Perhaps the process was too tricky to control, the raw materials too esoteric and expensive for common use, or the commercial outlets too few, or maybe they were not made aware of the advance that had been made. For whatever reason, Sheffield waited another half century before Sorby and others introduced a comparable process. Faraday at least gained some notoriety from the episode: he was presented with a fine Damascus sword, and to this day the Royal Institution holds sample razor blades made from his steel. Some examples of his steel-making are shown in Fig. 30.

Nine years after his first venture into materials research, Faraday was again prevailed upon to undertake another project as a result of an outside contract. As he wrote to a colleague in 1827: 'I have already (and to a great extent for the Royal Institution) pledged myself to a very laborious and expensive series of experiments on glass'. The reason for the parenthesis is that then as now the Royal Institution was living financially from hand to mouth, and sources of external income were being sought. Through his position as President of the Royal Society, Humphry Davy had strong influence with a body set up jointly by that organisation and the Admiralty, called the Board of Longitude, one function of which was to commission studies arising from the technical requirements of the Navy. Thus it was that the Board determined on the need to improve the quality of glass used to make the lenses in astronomical telescopes, since navigation at sea depended for its precision on the reliability of the data recording positions of the stars in the

Fig. 30. Samples of steel made by Michael Faraday.

Nautical Almanac. An important factor limiting the data was the availability of large flaw-free boules of glass for lenses.

This was one of the first ever excursions of central government into the funding of science-based projects: Humphry Davy had played an important part in achieving this objective, and could well have been keen to ensure that some fraction of the work being sponsored should be awarded to the organisation of which he was Director. A further (and more subversive) slant on the episode is that the need in question may not have been so great after all, and that the true motive for devising the project might have been to secure continuing government support for applied research. In the 1820s the shadow of European war had lifted, leading Davy and his colleagues to fear that Whitehall might lose its carefully nurtured enthusiasm for putting money into science: analogies with the peace dividend are hard to avoid! Be that as it may, Faraday was handed the task which was, to a considerable degree, to ensure his salary for the next four years.

The glass in question was lead borosilicate, chosen for its high refractive index, and it was the lead that gave Faraday a succession of problems. To keep the melt as uncontaminated as possible he made crucibles out of platinum foil, folded up and crimped, with the same material as a stirrer. The glass remained beautifully clear, but from time to time the crucible mysteriously became perforated, so that the contents poured out on to the iron grate underneath. How did it happen? Faraday tackled the problem with a chemist's eye, and discovered a simple phenomenon of oxidation and reduction: PbO overflowing from the crucible had fallen on the grate and been reduced by the iron to lead. In the molten state the lead had alloyed with the platinum, perforating the foil. The solution was simple: put another piece of platinum over the grate underneath the crucible!

Yet further difficulties followed. He tried to scale up the preparation after he had optimised the conditions, but on 16 January 1829 the following despairing observation appears in his notebook: 'This morning opened the furnace; it was a great disappointment; it was bad... full of dark clouds, each made up of several parts arranged parallel to each other, something resembling a mackerel sky'. So what had gone wrong? Very systematically Faraday worked through all the possible sources of difficulty. At first he thought the dark particles were platinum, because he was familiar with the quite different appearance of iron particles. In fact he had been using platinum alloyed with gold, but he did not think that was the cause of the problem. Finally he traced the origin of the dark particles to a lack of oxygen in the atmosphere above the crucible, because he had covered it with a lid to prevent contamination. Could PbO be reduced under these circumstances? If so, platinum was not the cause of reduction because adding finely divided platinum to the melt had no effect.

In the end, Faraday tracked down the culprit: traces of carbon in the iron forming the grate underneath the crucible. In the atmosphere lacking oxygen, small amounts of carbon monoxide were forming, which reduced the PbO to lead. Carbonic acid (which was the name given to carbon dioxide at that time) was then converted back to carbon monoxide by the hot iron and, as Faraday wrote in graphic terms in his notebook, 'went back to the

Fig. 31. Boules of glass made by Michael Faraday.

glass to repeat its evil operation and produce more metallic lead. In this way it was that the glass became sullied by smoking clouds of lead'. Having got to the bottom of the problem, large pieces of clear glass could be produced, as the contract with the Board of Longitude required. Some examples of the glass produced by Faraday are shown in Fig. 31. Nevertheless, Faraday's enthusiasm for the project began to wane, and on 4 July 1831 he wrote to the Board asking to be released: 'I wish to lay the glass aside for a while, that I may enjoy the pleasure of working out my own thoughts on other subjects'. What other subjects we now know, for it was only a few weeks later, on 29 August, that his notebook records the seminal observation that changed the whole direction of his career in research, namely, the discovery of electromagnetic induction.

It has been fashionable to regard the years that Faraday spent on materials research as to a large degree wasted. It has even been said that by forcing him into the field the progress of basic science (so triumphantly extended by his efforts after 1830) was impeded, and the episode has been taken as an awful warning against holding back the imaginative thoughts of able scientists by confining them to a straightjacket of applied research done under contract. In the present era, such a view would, of course be grossly politically incorrect: I believe it would also be false. As well as honing his skills as a masterly experimenter, the two episodes I have described provide us with wonderful examples of investigative insight. Creative design of experiments, based on optimal use of the materials in hand (for example, the concentric clay pots used in the steel work); great persistence in systematically varying the conditions of the experiment to track down the cause of unexpected side effects (as in the work on glass) and meticulous observation of fine detail (the mackerel sky of lead particles), are all qualities that shine through the efforts of the Royal Institution's chemical assistant. It may be that these qualities were innate, but would be hard to refute the proposition that they were refined and concentrated by grappling with the complex and demanding world of materials science.

part 3

SOME FOLKS YOU MEET

That science is carried out by scientists appears to be a truism so strikingly banal as to be hardly worth writing down. But the spirit of our age, at least among politicians and those who form 'policy' for science appears increasingly to regard new knowledge and insights as commodities to be ordered up and delivered, on-time and on-budget, against milestones and performance indicators. In a recent research proposal to the UK Engineering and Physical Sciences Research Council, I was sufficiently irritated by this tendency to write that, since we were proposing to make new molecules and new solids rather than roads and bridges, milestones were an inappropriate measure of progress. Still, thank goodness, reality keeps breaking through to confound the managerial jargon.

Approaches to finding new facts and theories are as diverse as human personalities—and perhaps (who knows?) even in what domain of knowledge an individual chooses to look for them may have psychological as much as professional origins. Long scholarly debates have taken place about the possible role, conscious or unconscious, that his deeply felt religion might have had on Faraday's to seek connections between what he called the 'various forces of nature'. And Rumford's lifelong preoccupation with the nature and manifestations of heat undoubtedly owed more than a little to his pragmatic and practical temperament.

The three 'early fathers' (to use the theological phrase), part of whose story was told in the previous section, were about as different in character from each other as it is possible for men to be. The same goes, too, for the three whose lives and scientific careers are memorialised below. Apart from the fact that they each made seminal contributions to their highly distinct disciplines, and that they died within the last few years, the only feature they have in common (just as for the institutions described in Section 1) is that I came to know them all well.

Respect (even, in present day street jargon, without the final 't') is not accorded lightly among scientists but each of these had it in spades, though in very different ways. Fred Dainton loved the House of Lords and the 'corridors of power' while Klixbull Jørgensen would have had quizzical words to say about Lords and masters; Olivier Kahn rallied his intellectual forces by fiery oratory. So, a Parisian, an expatriate Dane and a Yorkshire-man: a tiny sampling of the throng of distinguished individuals who 'made a difference', and of the thesis that it is scientists who make science, just as much at the end of the twentieth century as at the beginning of the nineteenth.

chapter 7

Christian Klixbull Jørgensen (1931–2001)

Inorganic Spectroscopist Extraordinaire

By devising simple models for correlating electronic spectra of inorganic compounds and through his inimitable style in expounding them, Christian Klixbull Jørgensen, who died on 9th January 2001, exerted a powerful influence on the development of inorganic chemistry from the 1950s on, and which continues (often unacknowledged) up to the present day. The style was the man, and the style consisted of recording and assimilating enormous quantities of experimental data, much obtained himself (he rarely had students or co-workers) and the remainder garnered from endless intense reading in the frequently obscure scientific literature. Indeed, the image of him that stays most readily in the mind is of a large domed head bent low over a tome in the library, for he was decidedly short sighted. It is, too, the archetypal image of a savant, for that he undoubtedly was, as well as, perhaps less kindly, a mad professor.

The making of a spectroscopist

It seems right to begin this brief appreciation of the life and work of Klixbull (as he was known to all but his closest friends and colleagues) on such a personal note, not only because it was my privilege to know and work with him over many years, but because, in a age where scientific knowledge is increasingly regarded as a commodity and its practitioners as ciphers, Klixbull was notably individualistic in his approach to the search for order in nature. He was born at Allborg, in the Danish province of Jutland on 18th April 1931. His father was an officer on a training ship but the family soon moved to Copenhagen, where he went to school and to university. He was a precocious pupil, with an early interest in science fostered by experiments at home with a chemistry set. It was there that began his lifelong fascination with the rare earth elements, which he tried to separate by fractional crystallisation. His interest in optical spectroscopy was also kindled by contact with a member of the staff of the Neils Bohr Laboratory in Copenhagen.

Arriving at the University of Copenhagen, he quickly extended the reading materials set by his teachers to include, somewhat surprisingly, the doctoral dissertation of Professor Jannik Bjerrum, who was later to become Klixbull's research supervisor. It is not hard to imagine the consternation of his already eminent professor on being relentlessly quizzed by a determined undergraduate about the details of his own thesis work. Perhaps it was Bjerrum's friendly response to this onslaught that persuaded Klixbull to take up graduate work in that laboratory. More likely, though, it was the combination of inorganic chemistry (which at that time was beginning to emerge from an empirical and fact-collecting phase into a search for more theoretically based explications) with the opportunity to exploit what was then an emerging new physical tool, visible absorption spectroscopy.

Visible absorption spectroscopy in the 1950s

In order to do justice to Klixbull's abiding influence on the development of models for chemical bonding in inorganic compounds (not only metal complexes) it is necessary to recreate the circumstances prevailing at that point in the evolution of the subject. Of course, in a general sense, the colours of inorganic compounds had been a source of interest to chemists for a very long time, often being the only physical property that could easily be reported, but the earliest optical spectroscopy concerned lanthanide compounds, in which the narrow intra-subshell transitions were susceptible to study by spectroscopes with photographic recording. Connecting these sharp spectroscopic features with chemical bonding we now know to be quite a subtle business, because to quite a good approximation the orbitals concerned do not play much of a role in bonding. However, following the Second World War it became much easier to record the broad and diffuse absorption spectra that most transition metal complexes show in solution at room temperature through the introduction of photocell detectors. This technical innovation also coincided with a surge of interest in the thermodynamics of forming metal complexes in solution and quantitative spectrophotometry, often at a single wavelength, when using a manual spectrophotometer such as the Beckman DU, came to the fore. Bjerrum's laboratory, being one of the world's leading centres for determining and rationalising complex formation constants, was equipped with the new technique and this, together with the young Klixbull's introduction to atomic spectra through his mentor in the Neils Bohr Institute, kindled the flame.

Methods and aims

Klixbull's very earliest publications were on thermodynamic properties in collaboration with his mentor and thesis advisor Jannik Bjerrum, but apart from one or two theoretical papers co-authored with Carl Ballhausen, practically all his publications (and there were no fewer than 44 in the first five years of his career) carry his name alone. As one who had the privilege of seeing him in action in the laboratory just a few years later, I can vouch personally for his prodigious rate of production. From the preparative laboratory (where quite often the product came directly from the supplier's bottle into the sample cuvette, to the spectrometer room where the Cary 14 ran more or less continuously, and on to the office, where the results were transcribed immediately and a manuscript written out in immaculate long-hand a few hours later, scarcely can the path from conception to publication have been shorter. The famous phrase attributed to Michael Faraday, 'work, finish, publish', applied precisely to Klixbull.

There was an exhilaration, too, in being at the opening of a new subject, when almost any new data, even on the simplest of compounds, raises questions and widens horizons. Of course, in this headlong rush mistakes were made—compounds wrongly identified, especially when multiple complexes were formed in solution or spectral bands overlapped. But the insights which moulded a discipline were not obtained by tireless study of the minutiae in single spectra, but through more or less empirical comparison of groups of related compounds: the classic inorganic chemist's approach, in fact. Fortunately, in the presence of such an avalanche of data, the approach was benignly fault tolerant and few conclusions from these years (and none of much consequence) have been overthrown by subsequent more careful mopping-up operations.

Despite the fact that Klixbull's approach to collecting spectroscopic data could appear rather cavalier at first sight, he was by no means neglectful of detail. In addition to correlating energies of spectral transitions, he was closely interested in features of the band profiles themselves, in particular intensities and line-shapes. Indeed, as early as 1955 we find him commenting on vibrational fine structure in one of the rare cases (actually $ReCl_6^{2-}$) where such a feature is seen in a room temperature solution spectrum. It is fair to say, though, that his attention was not captured strongly by such fine detail so that, for example, when it became possible to record polarised electronic spectra of metal complexes in crystals and hence apply symmetry selection rules, or to observe vibronic structure at low temperature, he did not deploy these techniques in his own work. Still less was he seized by the opportunities presented by magneto-optical spectroscopy (and in particular magnetic circular dichroism) from the early 1970s to define excited state wave-functions. Klixbull's driving purpose was to use a knowledge of the energy levels in inorganic compounds to derive models for their chemical bonding. The very first sentence of the Preface of his first book 'Absorption Spectra and Chemical Bonding in Complexes' proclaims his approach (just as the title itself does):

> *Our knowledge of the electron clouds of gaseous atoms and ions is based on the energy levels, studied in atomic spectroscopy. It has not been generally realised among chemists that the study of energy levels, i.e. the absorption spectra, is equally fruitful in helping our understanding of the chemical bonding, not only in transition group complexes, but in every type of molecule.*

Such a breadth of purpose led him, right from the outset, both to attempt correlation of spectroscopic parameters with other chemically relevant properties and to go beyond the simplest high symmetry metal complexes like the hexa-aquo-ions into lower symmetry and non-transition-metal compounds. For example, in 1956 he wrote on the influence of the spin-pairing energy and the consequent hump in the ionisation potentials of the 3d elements. He also wrote extensively about 4d and 5d complexes, and their striking differences from the corresponding 3d ones. At the same time, he was concerned, with Jannik Bjerrum, by what was then called crystal field stabilisation and the thermodynamics of complex formation in 3d elements.

Possibly as a result of his conditioning in early years by an interest in atomic spectra, especially of the 4f elements, throughout the 1950s Klixbull was greatly preoccupied by the role of electron repulsion in determining the transition energies in ligand field spectra. In particular, he drew attention to the fact that the values of the Racah–Cordon–Shortley parameters that gave the best fits to the observed spectra were invariably smaller than in the spectra of the corresponding gaseous ions. In a pseudo-atomic formalism, this implies that the average radius of the shell in question is greater in the coordination complex. Furthermore, for a given metal in a specific oxidation state, the effect varied with the coordinated ligands. With Claus Schaeffer, he placed all the common ligands in a series of increasing propensity to expand the metal d-shell and, after consulting a Professor of Classics at the University of Copenhagen, named it the nephelauxetic (Gk 'cloud expanding') series. In some ways that series, found in all undergraduate textbooks, is his most enduring monument. Earlier, in a nice piece of empirical ordering, he had shown that the spectrochemical series, which measures the ligand field splitting, could be factored into terms due to both the metal and the ligand. Now that a molecular orbital approach has long since taken over from crystal field or hybrid ligand field (let alone Pauling valence bond) models of coordination complexes, it is hard to recreate the confusion and controversy that resigned in this matter through the 1950s and 1960s. There can be no doubt, however, that Klixbull's insights, gained by correlating large numbers of spectra, greatly clarified the picture.

Charge transfer spectra

Harking back to the question above, which clearly announced the breadth of Klixbull's intentions as a spectroscopist, it is no surprise to learn that, in parallel with his immersion in what might appear to be the *minutiae* of electron repulsion in partly-filled d-shells, he was taking a wide view of molecular bonding by starting out on the first ever exploration into the detailed structure of charge transfer spectra. Looking back at the era in question, it seems somewhat strange that so much intellectual energy was being expended in understanding spectra that resulted from electric-dipole-forbidden transitions while at the same time the transitions that are fully allowed (and hence, in many cases, determine the colours of compounds themselves) were being strenuously ignored. All that can be said is that, at the time, there was a generalised perception that a correlation existed between dark colours and ready oxidation-reduction; assignment of the individual absorption bands to transitions between levels of defined (and verifiable) symmetry had never been attempted. In that context, Klixbull's 1959 paper on the charge transfer spectra of octahedral hexahalogeno-complexes of 4d and 5d elements represented a clear breakthrough. Not only did he lay out in an entirely convincing way how the bonding and non-bonding molecular orbitals of ligand type were ordered, but he assigned the complicated succession of absorption bands to transitions from these levels to the partly-filled d-shells. To do so, he invoked not only electron repulsion effects on multi-centre configurations (apparently for the first time) but also the influence of spin-orbit coupling on the ligands, laying a foundation for much future theoretical work, especially in photoelectron spectroscopy.

Yet it is entirely characteristic of Klixbull that he followed up this remarkable extension of quantitative theory into new territory, not by further refining the details, which was left to others using the novel techniques of low temperature high resolution spectroscopy and magnetic circular dichroism, but by using the results to derive a set of simple chemical parameters that could be applied to a wide range of compounds, giving insight to non-theorists and the general run of inorganic chemists. The concept of 'optical electronegativity' was founded on Mulliken's definition of electronegativity, but quantified using the electron transfer energies found in the absorption spectra, corrected for electron repulsion. For the halogens, for example, the resulting numbers mapped precisely on to the Pauling scale, but with the big advantage that parameters could be derived precisely, not just for elements but for individual oxidation states, and also for polyatomic groupings such as NCS. Later, in his book 'Orbitals in Atoms and Molecules' the concept was broadened to orbital electronegativity since analysis of the charge transfer spectra enabled the defining parameters to be determined orbital-by-orbital.

The angular overlap model

This brief survey of Klixbull's work has taken a more or less chronological approach rather than concentrating on the different classes of spectra that attracted his attention. Thus it is that, following his pioneering venture into charge transfer (or, as he always insisted on calling them, electron transfer) spectra, he returned to his earlier love, ligand field spectra, and embarked on a sustained effort to recast ligand field theory into a model comparable with the molecular orbital model that he believed provided the physically correct basis for rationalising the properties of partly-filled d- and f-shells.

One of the great strengths of the electrostatic crystal field model, taken over to a considerable degree by the more flexible ligand field approach, was its ability to treat low symmetry perturbations or mixed ligand environments by adding extra terms to the expansion of the central field around the metal in spherical harmonics. Values of the coefficients describing the contributions of these terms could be found by fitting the spectra, and interpreted through the chosen physical model of the central field. However, given that it was widely recognised by the 1960s that the orbitals concerned were not purely atomic, or even expanded atomic, but actually molecular orbitals containing a non-negligible ligand component, the question arose as to how this physical perception could be built on to the conventional model while retaining as much as possible its elegant simplicity. The result was the 'angular overlap model'.

Worked out and promulgated in close partnership with his long time friend and colleague Claus Schaeffer, the model took as its central simplifying hypothesis the idea that the group metal-ligand overlap integral associated with each symmetry-based molecular orbital could be factored into a product of a radial and an angular part. This was tantamount to assuming that the angle subtended by the ligand orbitals at the metal centre could be neglected—a surprisingly good and exceptionally useful approximation. Because of its relative simplicity in treating a very diverse range of problems that were directly interesting to inorganic chemists, the angular overlap model was taken up widely and enthusiastically. To this day, it continues to be a most helpful component in the coordination chemist's toolbox.

The Cyanamid Institute in Geneva

The body of work that we have just been surveying occupied essentially the decade of the 1960s. It was carried out, following Klixbull's departure from the University of Copenhagen, in the highly congenial and productive environment of the Cyanamid European Research Institute (CERI) just outside Geneva. This was such an unusual, and remarkably effective unit for generating new science, that it deserves a few words of description. During the later 1950s, a number of U.S. chemical companies found it both convenient and cost effective to open small basic research laboratories in Europe, for example Union Carbide in Brussels and Monsanto in Zurich in addition to CERI. The concept of CERI, set up by American Cyanamid Company in a refurbished and extended banker's villa on the shores of the Lac Leman, was especially innovative. The Institute was diverse but overall quite small: six groups, each of only three scientists, covering fields where the company's research management felt there could be future opportunities for product development. Thus, organo-phosphorus chemistry rubbed shoulders with crystal growth of new compound semiconductors, organometallic chemistry and organic electrochemistry. It is not quite clear what product area Klixbull's expertise matched, but most probably it was inorganic pigments and luminescent materials. In any case, looking back, the Institute was clearly intended at least in part as an advertisement for the company as an organisation operating at the boundaries of current science.

Klixbull's group was labelled 'theoretical inorganic chemistry', although much experimental work was undertaken, and coordination chemists of the older generation who found themselves on the CERI mailing list will recall the steady rain of envelopes through their mailboxes, bearing weighty preprints that carried the serial identification CERI-TIC followed by a rapidly rising number. In such diverse company, and with few administration constraints, Klixbull was in his element (Fig. 32). The Institute was a few miles outside the centre of Geneva, effectively in the countryside, so most members of staff stayed on-site at lunchtime to use the small cafeteria converted from what had been a gatekeeper's cottage. Such lunches were an invaluable and (at least to an impressionable young graduate student)

Fig. 32. Klixbull Jørgensen in the library of the Cyanamid Institute.

unforgettable forum for scientific discourse and speculation, starting with the soup and often lasting well into the afternoon. Whilst Klixbull's technique at a blackboard was to cover every inch of its surface systematically with formulae and sketched spectra (often in numerous colours), his custom of writing on the Formica-topped cafeteria tables with a pencil was broadly similar. Indeed, from time to time he found it necessary to migrate with the company between courses to another table, having filled the surface of the first one with diagrams.

But nemesis was at hand. In the late 1960s Cyanamid became involved in a legal battle that proved expensive for the company, so economies were sought. As is so often the case, long-range research was the first to suffer and CERI closed after an extraordinarily productive ten-year lifetime. Several senior scientists, including Klixbull, became Professors in the University of Geneva and began new careers as academics. For him, it was undoubtedly a scientific watershed.

The University of Geneva: Photoelectron spectroscopy, luminescence and other topics

Coincident with his new beginning in the University, Klixbull began to develop an interest in the then new field of photoelectron spectroscopy which, he felt, offered insights into chemical bonding in metal complexes analogous to, but potentially more extensive than, those furnished by visible-ultraviolet absorption spectroscopy 20 years earlier. Having purchased the equipment, he attacked the subject with the same vigour that he had brought to bear on ligand field spectroscopy, and with a similar approach. That is, he quickly recorded large numbers of spectra of sets of related compounds and began to correlate them. Data cascaded out, as evidenced by the title of one publication 'Photoelectron Spectra Induced by X-rays of Above 600 Non-metallic Compounds Containing 77 Elements'! Chemical shifts due to change of oxidation and fine structure due to electron repulsion of the core hole state and partly filled d-shell were identified. Unfortunately, however, there were problems. Certainly these were known to Klixbull, but perhaps were not treated by him with the seriousness that less imaginative and more pedantic experimenters brought to them. The first was surface charging. The coordination complexes and simple inorganic compounds of greatest interest to him were insulators, and while efforts were made to extrapolate results back to zero irradiation, doubts remained about precise energy calibrations. Another difficulty that makes photoelectron spectra more difficult to interpret than visible-ultraviolet spectra is the effect of the X-ray beam on the sample surface. Degradation is common and needs careful evaluation, only with difficulty compatible with the 'broad brush' approach preferred by Klixbull. Nevertheless, a large volume of publications attests to his efforts in this field; those that have had the greatest influence are probably the ones concerned more with concepts than data.

Towards the end of his research career Klixbull returned to his first love in optical spectroscopy, the lathanides and actinides, especially through a long and fruitful collaboration with Renata Reisfeld, concentrating especially on luminescence and energy transfer, the latter frequently connected with issues relating to solar energy conversion. The actinides (in particular the uranyl ion) preoccupied him, while a flash of his old enthusiasm for data collection can be discerned in an article on Mn^{II} emission in no fewer than

24 different phosphate glasses. The frontiers of his erudition continued to broaden, and he attempted to apply some of the principles of atomic spectroscopy that he had learned first many years before, including arguments based on symmetry, to the structure of nuclei and elementary particles. However, it is open to doubt whether any of these unconventional ideas found their way into the mainstream thinking of particle theory, although one at least found a place in the pages of *Nature*.

Throughout his life Klixbull had always taken a close interest in the history of science, not just as a further exercise in recondite fact finding, but as an essential tool for understanding how present concepts and approaches have developed and, above all, as a source for his ever present love of analogy. He was particularly delighted to draw attention to the discovery of Gd and Yb by Marignac in the laboratories of the University of Geneva, but in the title of his contribution on Marignac to the volume marking the century of the Geneva Chemistry Department, perhaps one may catch an echo of Klixbull's own situation: 'the solitary pioneer'. There can be no doubt that in the latter part of his life Klixbull, a naturally loquacious (indeed garrulous) and gregarious figure became increasingly isolated.

The early death of his wife Micheline in 1978 was not only a deep personal tragedy in itself but, for one not much concerned with everyday domestic matters, made life difficult. In the University, his unusual style and manner of working did not attract many graduate students, and while his pedagogical approach was rich in insights for those who were already familiar with the subjects he taught, less motivated students found his lectures too challenging to be easily accessible. In his books and articles, his arguments often developed through sequences of quite convoluted analogy, buttressed by dense arrays of references, frequently to obscure sources. The latter was the result of voracious reading and a remarkably retentive memory. Colleagues soon found that, when searching for a reference, it was often quicker to ask Klixbull for the volume and page number than to go to the library. In the end, perhaps it could be said that erudition was his greatest enemy in making contact with the wider community of scientists.

Klixbull's lasting memorials in science are the nephelauxetic series, the angular overlap model and optical electronegativity. To those who knew him, or saw him in action, however, the lasting impression is of one consumed by an urge to know; to gather, sift and synthesise facts and extract from them models of how the world works. Is there a better epitaph for a scientist?

'Whereof Man Cannot Speak'

Words for things

In their pursuit of larger truths, often encapsulated in short aphorisms, even the greatest philosophers can sometimes be guilty of making approximations. Thus, when Ludwig Wittgenstein finished his major work on linguistic philosophy *Tractatus Logico-Philosophicus* with the famous phrase 'whereof man cannot speak, thereof he must be silent', he failed to take account of a situation that occurs quite frequently in science, as newly uncovered facts about the world call for new concepts to describe them. In science, when we find something for which there is no existing vocabulary, it is certainly not customary to remain silent. What is normally done is to invent a new word or phrase.

This procedure of coining new phrases to embrace new facts or new levels of understanding goes back to the very foundations of the scientific enterprise. Indeed, a good case could be made that it is a fundamental part of the whole procedure of gaining intellectual ascendancy over the world of natural phenomena. Astronomers of the era of the scientific revolution were adventurous employers of vocabulary, and the chemists, having such a diverse array of phenomena to encompass, were not far behind: Just think of the potent sway that the concept symbolised by the word 'phlogiston' held over chemical theory up to the late eighteenth century. In the first half of the nineteenth century, one of the most prolific coiners of new phraseology was Michael Faraday. In fact, two quite disparate fields of science still bear the hallmark of his invention till the present day. Early in Faraday's career, electrochemistry (a word that we owe to Faraday's mentor, Humphry Davy) was at the centre of his interests, while two decades later it was magnetism. Before considering some of Klixbull Jørgensen's contributions to the language of chemistry, it is pertinent to document briefly some of the great contributions of the same kind by his illustrious forebear.

Michael Faraday and scientific vocabulary

One of many remarkable features of Faraday's career was his lack of advanced education. Indeed, his formal education ceased at the age of 13, in 1804, when he was apprenticed to a bookseller and bookbinder Mr Riebau. Consequently, it is not surprising that his knowledge of mathematics remained rudimentary throughout his life, probably not stretching much beyond simple arithmetic. Although his laboratory notes contain geometrical diagrams (for example the lines of induction around magnetised bodies), there is no evidence that he was capable of drawing quantitative conclusions from them. Among the earliest of many protestations found in his voluminous writings about a handicap in comprehending mathematical reasoning is a letter to Ampère in 1822, in which he takes refuge in the primacy of experimental facts, regarding mathematics as a navigational compass of uncertain efficacy. Here is what he wrote:

> *I am unfortunate in a want of mathematical knowledge and the power of entering with facility into abstract reasoning. I am obliged to feel my way by facts closely placed together, so that it often happens I am left behind in the progress of a branch of science (not merely from the want of attention) but from the incapability I lay under of following it, notwithstanding all my exertions. It is just now so, I am ashamed to say, with your refined researches on electromagnetism or electrodynamics. On reading your papers and letters, I have no difficulty in following the reasoning, but still at last I seem to want something more on which to steady the conclusions. I fancy the habit I got into of attending too closely to experiment has somewhat fettered my powers of reasoning, and chains me down, and I cannot help now and then comparing myself to a timid ignorant navigator who (though he might boldly and safely steer across a bay or an ocean by the aid of a compass which in its actions and principles is infallible) is afraid to leave sight of the shore because he understands not the power of the instrument that is to guide him.*

In contrast to his suspicion of mathematics as a tool, Faraday was intensely concerned with inventing and defining new words to encapsulate the new phenomena that his experiments were opening up. The ironic tone of the following description (in a letter of 1831 to the chemist Richard Phillips) of how the word 'electrotonic' (now quite vanished from the scientific vocabulary) came to be coined, should not be taken at face value:

> *This condition of matter I have dignified by the term Electro-tonic. What do you think of that? Am I not a bold man, ignorant as I am, to coin words? But I have consulted the scholars.*

Despite his own propensity for coining words, Faraday was not above casting scorn on others who sought to do the same. The line between illumination and obfuscation in the way that technical language is used remains a narrow one for science even today. William Whewell (the Cambridge historian and classical scholar) frequently gave advice to Faraday about etymology, enveloping the new objects of science in classical language. In one of his letters, Faraday wrote to Whewell as follows:

> *Your remarks upon chemical notation, with the variety of systems which have arisen with regard to notation, nomenclature, scales of proportional or atomic number etc., etc., had almost stirred me up to regret publicly that such hindrances to the progress of science should exist. I cannot help thinking it a most unfortunate thing that men who, as experimentalists and philosophers, are the most fitted to advance the general cause of science and knowledge should, by the promulgation of their own theoretical views under the form of nomenclature, notation or scale, actually retard its progress.*

Faraday's earliest lasting contribution to scientific nomenclature arose from his experiments in the 1820s and early 1830s on the electrolysis of aqueous solutions. He also sought advice about nomenclature from another friend, William Nicholl, which he passed on in a letter to William Whewell in 1834. That letter contains the first ever mention of the words 'electrode' and 'electrolyte'. In contrast, it is interesting to observe that another word 'zetode' did not survive. In his letter to Whewell, he writes:

> *I wanted some new names to express my facts in electrical science without involving more theory than I could help, and applied to a friend Dr Nicholl, who has given me some that I intend to adopt. For*

instance, a body decomposable by the passage of the electric current, I call an 'electrolyte', and instead of saying that water is electrochemically decomposed I say it is 'electrolysed'. The intensity above which a body is decomposed, beneath which it conducts without decomposition, I call the 'electrolytic intensity', etc., etc. What have been called the poles of the battery I call the 'electrodes'. They are not merely surfaces of metal, but even of water and air, to which the term poles could hardly apply without receiving a new sense. Electrolytes must consist of two parts which, during the electrolysation, are determined the one in one direction, the other in the other towards the electrodes or poles where they are evolved. These evolved substances I call 'zetodes', which are therefore the direct constituents of electrolytes.

Faraday also accepted other suggestions from Whewell, amongst which anode, cathode, cation and anion are still with us, and furnish the staple vocabulary of electrochemistry, although others (dexiode and skaiode) did not stand the test of time. In the following letter thanking Whewell for his help, he is clearly exercised not only about the clarity for the new words but also by their euphony:

All your names I and my friend approve of (or nearly all) as to sense and expression, but I am frightened by their length and sound when compounded. As you will see, I have taken dexiode and skaiode because they agree best with my natural standard East and West. I like Anode and Cathode better as to sound, but all to whom I have shewn them have supposed at first by Anode I meant No way.

I have taken your advice, and the names used are anode, cathode, anions, cations and ions; the last I shall have but little occasion for. [How wrong he was in that prediction!] *I had some hot objections made to them here and found myself very much in the condition of the man with his son and ass who tried to please everybody; but when I held up the shield of your authority, it was wonderful to observe how the tone of objection melted away.*

Twenty years on from his electrochemical experiments, Faraday was working on magnetism. This period gave us two more words, used in the physical sciences to this day, diamagnetism and paramagnetism. The concept that lines of magnetic induction are more closely packed together inside a paramagnetic body than in free space, and less so in a diamagnetic one, remains valid after 150 years. Again, he reported his choice of words in a letter to Whewell, at the same time hazarding the idea that paramagnetic oxygen in the atmosphere might contribute to the earth's magnetism (now we know that it does not). It should also be noted how defensive he sounds about sharing his ideas until they are fully worked out for publication—a thought that resonates strongly with the modern world! The 1850 letter to Whewell reads as follows:

> *I have been driven to assume for a time, especially in relation to the gases, a sort of conducting power for magnetism. Mere space is Zero. One substance being made to occupy a given portion of space will cause more lines of force to pass through that space than before, and another substance will cause less to pass. The former I now call paramagnetic and the latter are the diamagnetic. The former need not of necessity assume a polarity of particles such as iron has when magnetic, and the latter do not assume any such polarity either direct or reverse. I do not say more to you just now because my own thoughts are only in the act of formation, but this I may say: that the atmosphere has an extraordinary magnetic constitution, and I hope and expect to find in it the cause of the annual and diurnal variations, but keep this to yourself until I have time to see what harvest will spring from my growing ideas.*

My reason for documenting these thoughts at this point is not in any way to belittle the accomplishments of Klixbull Jørgensen in the much narrower field of science in which he operated, but rather to illustrate how important it is to enlarge language to symbolise new concepts. Before describing some of Jørgensen's creative flights with language, however, it is pertinent to recall a few facts about his origins.

Klixbull Jørgensen and the Language of Science

Beginnings

In almost no respect did Jørgensen's upbringing and education resemble that of Faraday, save that both were precocious students. Jørgensen's father was an officer on a naval training ship but the family moved to Copenhagen, where he went to school and to university. He was a voracious pupil, with an early interest in science fostered by experiments at home with a chemistry set, much as in Faraday's case. It was there that he began his lifelong interest in the lanthanides, which he tried to separate by fractional crystallisation. His interest in optical spectroscopy was also kindled by a member of the staff of the Niels Bohr Laboratory in Copenhagen. At the University of Copenhagen he quickly moved beyond the reading set by his lecturers to include, for example, the doctoral dissertation of the then Head of Department, Professor Jannik Bjerrum, who later supervised Jørgensen's research. One can imagine how surprised the eminent professor must have been at being interrogated by one of his students about the details of his own thesis.

Jørgensen's first publications were with Bjerrum, but practically all the publications from the early part of his career (and, indeed, many subsequently) were his alone. In this respect at least, his career does resemble that of Faraday, who worked and published mostly alone. A further point of resemblance is in speed of working: in the first five years of his career, Jørgensen published 44 papers. Of course, the compass of Jørgensen's interest was narrower than Faraday's, but it is certainly fair to say that, at the earliest part of both men's careers, they had the good fortune (or, more probably, good judgement) to find themselves at the point of entering into a new chemical landscape; in Faraday's case it was electrochemistry, in Jørgensen's, chemical spectroscopy. In intellectual terms, both fields were at that early stage where practically any new observation opens a new horizon for speculation. That suited Jørgensen's style of working very well: he was in the fortunate position that the then newly developed technology of photoelectric detectors made it possible for the first time to record visible-ultraviolet spectra of metal complexes in solution, where the absorption bands were often weak and rather broad. Almost any compound taken from the bottle and dissolved in water would have a spectrum exhibiting some new feature. More important than that, the technology made it possible quickly to build up series of spectra of closely related compounds and hence establish correlations of the kind that form the traditional methodology of inorganic chemistry.

In the hands of the unimaginative, that may be called 'stamp collecting' but, as a prelude to comprehending the manifold factors contributing to a given outcome, it is an essential first step. And Jørgensen, like the young Faraday, was committed from the start not just to interrogate nature but to comprehend it. That, of course, meant theory, but the theory of electronic structure of many electron systems like transition metals and their complexes was not then (or even now) susceptible to exact solution. Judgements as to approximations were therefore fundamental to the process of extracting the maximum of insight with the minimum of complexity. Jørgensen's own view of the place of theory in chemistry is expressed in characteristically graphic terms in the following:

Chemistry is very young as a science, and the most plausible and coherent hypotheses about the nature of chemical bonding have changed very rapidly. However, the waves of new discoveries have deposited a sediment of theory, which is not so thixotropic as sand though it is not as firm as rock.

So it was in pursuit of understanding chemical bonding by means of informed approximation that Jørgensen embarked on his relentless accumulation of spectroscopic data. In this he proclaims his purpose, in the very first sentence of the Preface to his first book 'Absorption Spectra and Chemical Bonding in Complexes' (as, indeed, also in the title):

Our knowledge of the electron clouds of gaseous atoms and ions is based on the energy levels, studied in atomic spectroscopy. It has been generally realised among chemists that the study of energy levels, i.e. the absorption spectra, is equally fruitful in helping our understanding of the chemical bonding, not only in transition group complexes, but in every type of molecule.

We find the same sentiment at the end of his fascinating book 'Oxidation Numbers and Oxidation States':

Since the properties of matter surrounding us are essentially dependent of the electronic structure alone, the combination of chemistry and spectroscopy is of utmost importance for our preliminary and provisional, but ever progressing understanding of our complicated and fascinating world.

Let us now examine a few of the new concepts and the imaginative new terminology that Jørgensen coined to describe them.

Insight from neologisms

Preponderant configurations and ionic colours

Jørgensen was never much taken with the minutiae of optical spectroscopy; polarisation selection rules, cryogenics, magneto-optics and so on concerned him little. The energies and intensities of transitions were sufficient for his purposes of identifying what he called preponderant configurations. Indeed, completely qualitative observations were sometimes brought to bear, as in his coining of the phrase 'additivity of ionic colours'. By that he meant cases where, for example, a blue cationic complex formed a salt with a yellow anionic one, creating a green solid and indicating that not only did the complexes retain their electronic integrity in the new environment, but that there was no significant electronic interaction between them. Conversely, where there was non-additivity, such interaction existed and was manifested either in substantial spectral shifts or the appearance of new absorption bands. A famous case of the former is Magnus' Green Salt, $Pt(NH_3)_4PtCl_4$, formed from a colourless cation and a pink anion and of the latter the many instances of inter-valence transitions in mixed valency compounds.

The nephelauxetic effect

Of all the words and phrases invented or applied by Jørgensen to encapsulate the phenomena that he was uncovering in the electronic spectra of inorganic complexes, the one which is likely to prove the most durable is 'nephelauxetic'. In a few favourable cases the energy of a particular ligand field transition is determined only by a variation in inter-electron repulsion, rather than orbital excitation. Such transitions are called intra-configurational since they correspond to a change in the total spin quantum number within a given configuration. Examples are:

$$^{6}A_{1g} \rightarrow {}^{4}A_{1g}, {}^{4}E_{g} \text{ in octahedral Mn}^{II}; \text{ and}$$

$$^{4}A_{2g} \rightarrow {}^{2}E_{g}, {}^{2}T_{1g} \text{ in octahedral Cr}^{III}.$$

Since there are very large numbers of each kind, and the transitions are relatively easy to identify in the spectra, placing them in an empirical series is straightforward. In fact, the corresponding effect in 4f complexes had been known for half a century, and some discussion of the difference between hex-aquo- and hex-ammino-complexes had already taken place. Adding a much larger number of ligands to the series, however, served to emphasise a qualitative correlation with their reducing character and, since the transition energy depends only on the inter-electron repulsion, served to emphasise change in the effective radius of the 3d orbital going from the gaseous ion to the complex. The more reducing the ligand, the lower the repulsion and the larger the effective radius.

Just as Faraday approached Whewell for words to add classical dignity to his new results, so Jørgensen turned to Professor K. Barr of the University of Copenhagen. What emerged was 'nephelauxetic' from the Greek for 'cloud expanding'. It was further noted that the nephelauxetic ratio β (being the ratio of the electron repulsion parameter B or C found in the complex to that of the gaseous ion in the same oxidation state, i.e. the same preponderant configuration) was determined by the interplay of two phenomena, to which Jørgensen gave the further labels 'central field' and 'symmetry restricted' covalency. The former implies the invasion (Jørgensen's word) of the central ion by electrons from the ligand forming bonding molecular orbitals and hence decreasing its effective charge, while the latter arises from the contribution of the ligand atomic orbitals to the anti-bonding molecular orbital containing the unpaired electrons. Somewhat later Schaeffer proposed the ratio of ligand field splitting to electron repulsion (Δ/B) as a measure of ligand field strength, to which he assigned the symbol Σ. This brought forth from Jørgensen a further linguistic flourish. He pointed out that the latter is the first letter of the Greek work σφοδρότης, meaning 'force applied against a resistance', in this case inter-electron repulsion acting against spin-pairing. Correspondingly, Δ can be thought of as the first letter of δυωαμις, meaning force as a potential ability. Finally (even more whimsically), the letter β might easily have been chosen for the nephelauxetic ratio for the word βάτραχος (frog) from the legend about the frog wanting to blow himself up to the size of an ox! Unfortunately, what Jørgensen calls 'this mnemotechnic derivation' is, as he writes, *a posteriori*.

Innocent ligands

Other well judged epithets for phenomena connected with the distribution of charge and its rearrangement after electronic excitation flowed from Jørgensen's copious imagination. Thus, in an adjective still widely used, he distinguished ligands like H_2O with closed shells, making complexes with unambiguously defined oxidation states, from ones like NO and maleonitrile-dithiolate, where the degree of electron donation is uncertain, by calling the former 'innocent'. Sadly he never labelled the antithesis—would it have been guilty? Where more than one non-innocent ligand is attached to a central metal atom, he also found it helpful to speak of the ligand array as being 'collectively oxidised'. In the excited state of an electron transfer transition, such as that of $IrCl_6^{2-}$, where an electron has been transferred from a ligand-based molecular orbital to the metal 5d shell, the same situation applies, i.e. to a good approximation we have $(Cl)_6^{5-}$. Likewise, the tendency of hard or non-polarisable ligands to come together around a metal centre, or of soft with soft, he correctly called 'symbiosis', extending the meaning of the word beyond its usual biological context of two species interdependent on one another. Thus, for example, $Mn(CO)_5X$ are far more stable for $X = I$, H, CH_3 and have not been synthesised for $X = F$.

Oxidation numbers and oxidation states

The phrase 'oxidation state' goes right to the heart of inorganic chemistry, and was current for very many years before Jørgensen came on the scene. There is no doubt though that it has been defined in a variety of ways and hence employed in quite an ambiguous fashion. Jørgensen applied himself to setting it in the context of modern theories of chemical bonding, in particular in one of his most eloquent and insightful books 'Oxidation Numbers and Oxidation States' (already referred to above), which provides us with further examples of his analytical and expository style. In the course of the first few pages of text he introduces no fewer than five different concepts related to oxidation. First are formal oxidation numbers, to be written as superscript small Roman numerals such as Cu^{II}, Cr^{III} though with the anachronism that zero and negatives can be added. These are to be distinguished from (spectroscopic) oxidation states, which are derived from the excited states observed in the visible-ultraviolet spectrum. In that case the Roman figures are written on the line and in parenthesis, such as Cr(III) and may also occur in the description of chromophores like $Cr(III)O_6$, occurring in $KCr(SO_4)_2 12H_2O$ and Cr_2O_3. Next came conditional oxidation states, defined with respect to the parent configuration found from atomic spectroscopy, but taking account only of the partly filled shell having the smallest ionic radius. Thus the three configurations of Ti^{2+}, Ti^+ and Ti^0 (respectively $[Ar]3d^2$, $[Ar]3d^24s$ and $[Ar]3d^24s^2$) would all be counted as being Ti[II] (note the square brackets). One reason for this classification is the simultaneous occurrence of localised and delocalised electrons in some solids, e.g. metallic PrI_2, which would be counted as Pr[III] from its magnetic properties, or even the lanthanide elements themselves. It also makes contact with the concept of preponderant configuration mentioned earlier.

The two final definitions, quanticle oxidation states and distributed quanticle oxidation states, take us into more controversial territory. The word 'quanticle' was coined by Fajans to denote one or more (most frequently an even number) of electrons quantised with respect to one or more nuclei. Where only one nucleus is concerned, of course this is synonymous with nl-configurations, but where there are more, it can be taken as representing the molecular orbitals of the valence shell. Jørgensen writes the atomic constituent lacking all the electrons taking part in the bonding quanticles as e.g. N{III} in NH_3 and N{V} in NH_4^+, while the 'distributed' variant is obtained by sharing each two-electron quanticle equally between the two atom bonds, e.g. N{0} in NH_3 and NF_3, $N\langle 1\rangle$ in NH_4^+.

Apart from his having accepted a contract to write a book on 'Oxidation Numbers and Oxidation States', and a decision to enter fully into all the ways in which such terms could be defined, it is hard to imagine from our current perspective why the author should have devoted such intellectual energy to these arcane distinctions. Quanticles have certainly taken the same path to oblivion as Faraday's zetode. But it is nevertheless fascinating to observe the performance and the manner of deploying a formidable body of factual information and theoretical insight. As he acknowledges, 'there seems to be a kind of inverse variation of the physical content of various possible definitions of oxidation states and the number of compounds to which these definitions can be applied'.

Chemical taxonomy and taxological quantum chemistry

As we have seen in earlier sections of this appreciation, two themes run through Jørgensen's approach to chemical theory. These are, first, a reliance on spectroscopic evidence to establish 'preponderant configurations', naturally of the excited states themselves, but by extension, also the ground state. By 'spectroscopy' we mean, in the first instance, excitations within and around the valence shell (frontier orbitals in the organic chemist's phraseology) observed by electronic absorption and emission spectra up to (say) about 6 eV. Towards the end of his career as an experimenter, Jørgensen turned to the then newly invented photoelectron spectroscopy in an effort to broaden the horizon of energy levels being brought into consideration, but his results are less illuminating, in part because of experimental difficulties in respect of surface damage and charging, so they are not referred to here. The second impulse driving his science was the desire to comprehend, through classification, what were the major factors governing not just the appearance of the spectra that he empathised with so closely, but the quantum chemical bonding principles that determine them. In this, he aligned himself with the traditional approach of the inorganic chemist (or, as he once described himself to the author, 'inorganic physicist'):

> *... if people still devote their efforts to try to understand chemistry, there is hope. The empirical facts of chemistry have been fashioned into many theoretical descriptions with time; most of these descriptions have been abandoned again; but there remain always some pleasant memories.*

Classification means labelling, and labelling often means finding new words. The eighteenth and nineteenth centuries represented the apogee of taxonomy, when biologists imposed order on a vast array of newly found species by inventing botanical and zoological appellations. Jørgensen had a similar conception of chemistry beginning as natural history but progressing to understanding through classification, and introduced the new word taxology, which he defined as follows:

> *When we use the word taxology about quantum chemistry, we are trying to draw attention to the meta-theoretical, the higher-type aspect in Bertrand Russell's sense, propensity of preponderant electron configurations suitably chosen to classify correctly the symmetry types of the ground state and lowest excited states.... No criterion is safer for the presence of a definitive spectroscopic oxidation state than the observation of such a complicated manifold of excited levels with the predicted symmetry types. The paradoxical situation is that this classification works even though we know that Ψ of many electron systems do not correspond to well-defined configurations.... The whole theory of such configurations is a masquerade played by Nature; it is as if the preponderant configurations are taxologically valid.*

The mention of Bertrand Russell in the above quotation is representative of a deeper interest in the philosophy of classification, which emerges in the following:

> *Whereas Aristotle was interested both in formal logic and in botany, there has not in recent times been a very strong interaction between chemistry and formal logic... the British philosopher Jevons pointed out that the natural sciences in actual practice frequently use definitions of the type 'metals are materials having the properties common to sodium, iron and gold'... the examples are chosen as different as possible and yet compatible with the intention .... Both Aristotle and the reflecting chemist immediately start wondering about essential and accidental properties.*

What is interesting here is that Jørgensen, good chemist as he was, makes no mention of the simple theory-based definition that a metal is a solid in which one can define a Fermi surface. But the problem can be posed when defining oxidation states from preponderant electron configurations, and Jørgensen comes back again to Bertrand Russell's theory of types, i.e. that it is the class itself and not its members individually which may be numerous. Thus the ground and lower lying excited states of IrF_6 all belong to the situation where three 5d-like electrons and hence the oxidation state Ir^{VI} is a good descriptor. Without committing oneself to any judgement about the fractional charge of the central atom (probably between 2 and 3), higher energy excited states belong to a preponderant configuration with four 5d-like electrons and hence designated Ir^{V}, probably with a similar fractional charge on the Ir.

As an amusing aside on essential and accidental properties, Jørgensen invites us to consider the class of halogens: it is certainly a chemically interesting statement that all halogens have odd atomic numbers. Given the laws of atomic structure, it is inconceivable that any newly discovered halogen would not. On the other hand, it is certainly an accident (as Jørgensen says, only of interest to someone having access to just the first volume of the Chemical Abstracts index) that the English names of all the halogens begin with one of the first ten letters of the alphabet! The key issue regarding the validity of definitions or classifications is thus their stability against future discoveries. Astatine could equally well have been called syntomine (from the Greek 'brief') or taxyboline (from the Greek 'rapid gunfire') according to Jørgensen, but the point about atomic number remains robust.

Chemistry, philosophy and language

Although the phrase 'natural philosophy' has a long and honourable pedigree, the words 'chemical' and 'philosophy' are not associated very much. In part, perhaps that is because chemistry (at least in earlier times) had more the character of natural history, that is, of a dense and fascinating jungle of compounds and phenomena, through the trees of which it was often quite difficult to see the wood. Even today, it remains notoriously hard to predict the crystal structure of a new molecular compound from a knowledge of its chemical formula. Chemistry, therefore, remains wonderfully capable of springing surprises (fullerenes and molecular superconductors, to name but two relatively recent examples). Or, as Jørgensen wrote in the late 1960s, 'at one time, the class of non-metallic ferromagnets was thought to be as empty as earlier the class of aerostats heavier than air'. So description and classification founded on quantum-mechanically-based approximation remains a viable (indeed, in many cases the only) option. A final quotation:

> *It is not reasonable for the chemist to sit down and wait for the pure theorists to deduce everything ab initio; experience shows clearly that the preliminary steps of approximate calculation have an unfortunate tendency of producing nonsensical results from incorrect assumptions, unless these are strictly governed by a realistic knowledge of the experimental facts.*

Classification and description need words, both the facility to manipulate the existing ones and an ability to coin new ones. I have tried to show in this brief memoir that Jørgensen was a master of language, albeit on occasions a somewhat idiosyncratic one. At no point up till now has any reference been made to the fact that English was not his first language. Like so many of his compatriots, however, he was equally fluent in several languages, though with a pronunciation that remained resolutely Danish. Access to the instincts of vocabulary in a more general sense must be a prerequisite to coining new words and phrases for scientific purposes. A glance at Faraday's voluminous writings shows that he too had that instinct for language. There is no doubt that both men enriched our scientific vocabulary greatly while searching to encompass their new findings.

chapter 8

Olivier Kahn (1943–1999)

A (too) Brief Life

Olivier Kahn (Fig. 33), a pioneer and protagonist of molecular-based magnetic materials, and a major influence on contemporary inorganic chemistry, died suddenly in Paris on 8 December 1999 of an aneurysm at the early age of 57. He had just returned from a visit to Japan, where he was collaborating with a multidisciplinary group at the Institute of Physical and Chemical Research (RIKEN), and the following day was due to chair a meeting of the French Chemical Society on coordination chemistry. These are just two examples, in his always-crowded diary, of the international reach of his interests and his intellectual roots as an inorganic coordination chemist.

Over the preceding 20 years, Olivier Kahn had played a key part in transforming inorganic chemistry from a science largely concerned with individual molecules to one in which greater emphasis is placed on the behaviour of bulk material. He was born and grew up in Paris in a family in which intellectual debate, in the highest tradition of French metropolitan life, was part of the atmosphere (one of his two brothers is a noted social and political commentator on French television, the other an eminent molecular biologist). Graduating first in his class in chemistry from the École Nationale Supérieure de Chimie, he carried out his doctoral research at the same school. Particularly influential in developing his ideas was a postdoctoral stay in Britain, at the University of East Anglia, in the early 1970s, where he

Fig. 33. Olivier Kahn.

learned about the quantum-mechanical approach to magnetic interactions from the theoretical chemist Sid Kettle.

The 1950s and 1960s had witnessed a renaissance in inorganic chemistry, driven by attempts first to synthesize and then to rationalize the structures and properties of new compounds in which metal atoms (usually from the transition elements) were surrounded by organic molecules called ligands that modulate both structure and reactivity. At the time, the dominant theoretical framework for understanding the properties of transition metal compounds was ligand field theory, an idea going back to Bethe's electrostatic approach of the 1920s. This theory necessarily focussed attention on single metal atoms and their immediate surroundings (usually organic ligands)—a mindset that became even more entrenched with the explosive flowering of organometallic chemistry. Cracks were already beginning to appear in this purely molecular view of inorganic chemistry. They were widened in part by a resurgence of interest in the classical solid-state chemistry of oxides and other continuous lattice compounds, pioneered by J. S. Anderson and others. It was also beginning to be appreciated how important interactions between molecules can be in determining properties of matter in bulk, such as cooperative magnetic and optical phenomena in one- or two-dimensional solids. Into this epoch burst the figure of Olivier Kahn.

Kahn grasped the opportunity presented by his appointment as the youngest full professor at the University of Paris-Sud, Orsay, to build up (and, in the expressive Gallic phrase, 'animate') a new laboratory. He was eager, following his apprenticeship in coordination chemistry and molecular quantum mechanics, to create and articulate a new subject. Early success came with the design and synthesis of a molecular complex containing one element from the beginning (vanadium) and one from the end (copper) of the first transition series, placed side by side so that orbitals containing the unpaired electrons were strictly orthogonal to one another. This arrangement was predicted by P. W. Anderson, M. Kanamori and J. B. Goodenough to lead to a parallel alignment of the spins, resulting in ferromagnetism (that is, each molecule becomes strongly magnetized, like permanent magnets such as iron). It was a triumph. A ferromagnetic ground state was observed, though not of course one where the ordering was long range: this was molecular ferromagnetism.

In quick succession, Kahn made further marks on coordination chemistry by two contrasting demarches. First, with B. Briat and others, he recast the Anderson–Kanamori–Goodenough model of exchange interactions between localized magnetic moments into a form accessible to chemists. Second, he extended his synthetic success with ferromagnetic-dimers to infinite chains of transition-metal ions coupled through organic bridges to make ferromagnetic arrays, in which the spins are aligned anti-parallel, but unequal contributions from each ion still produce magnetization. In this way, Kahn connected the molecular world with the physicists' paradigm of infinite one-dimensional magnetic order.

Magnetic chains may be infinite in one sense, but are not themselves true bulk magnets because the chains must interact with each other to give three-dimensional order. It was then that Kahn showed, by a beautiful piece of supra-molecular chemical tweaking, how the chains in his one-dimensional ferrimagnets could be arranged so as to give ferromagnetic interactions between the chains. Other lattice topologies, followed, including one with three-dimensionally interlocking rings of metal atoms joined together by organic ligands (Fig. 34). This one Kahn compared, with a characteristic flight of fancy, to a necklace that he had bought for his wife.

Throughout his career, first at Orsay and later Bordeaux, where he established a large and flourishing laboratory in the 1990s, Kahn led his young team by his own lively example: exclamations of delight at the latest findings of a graduate student echoed through the corridors. When a colleague chided him gently for spending little time at home, his reply was 'but my wife and I are always together on Sunday afternoons'.

For Olivier Kahn, finding beautiful new patterns in nature—whether strung together by necklaces of atoms in a crystal lattice, or in his more recent theories of molecular bi-stability as a basis for information storage—was an all-consuming passion, which he shared ebulliently with everyone he met. For his friends across the globe, as well as for the molecular sciences he espoused so eloquently, the hole left by his passing will be hard to fill. Not to see that eager figure, bolt upright in the centre of the front row in the lecture theatre, spectacles gleaming, poised to ask the first question, makes science duller as well as sadder.

Molecules and Magnets: The Legacy of Olivier Kahn

Progress and people

The approach to history symbolised by the phrase 'Cleopatra's nose' has never been an attractive starting point for understanding how science progresses. The notion that, had Cleopatra's nose been half a centimetre shorter or longer, world history would have been different, hardly seems to apply to an endeavour that appears to consist in uncovering objective laws of nature, which presumably would be the same irrespective of who first discovered them. Yet we all know that the subject matter of science cannot be entirely divorced from those who practise it, and that scientific enquiry sometimes takes particular directions as a result of the interest or advocacy of particular individuals. To exercise such a potent influence on scientific thinking requires three unusual combinations of factors to be brought together. The first is a combination of background knowledge from fields not previously thought of as having much in common; the second is a combination of diverse talents to bring the new perspective to the notice of others, and the third is a combination of circumstances at a particular moment in the evolution of a scientific discipline (perhaps a kind of '*zeitgeist*') when conventional approaches are seen to be inadequate in the face of new knowledge. All three of these coincided in the case of Olivier Kahn.

First, Olivier Kahn combined the background of a preparative coordination chemist with a deep knowledge of the theory of chemical bonding; second he was a fine communicator with highly developed histrionic skills, and an enthusiastic invigorator of talented young people; third (and perhaps most important), he came to his intellectual maturity at a key moment in the development of inorganic chemistry as applied to condensed matter.

Inorganic chemistry: Molecules and solids

Although the classical solid state chemistry of oxides, halides and pnicnides formed an important part of inorganic chemistry in the earlier part of the present century, by the 1940s and 50s attention had largely shifted away from the continuous lattice solid state towards the chemistry of coordination complexes and, later, organometallic compounds. Whether charged or neutral, the fundamental characteristic of these components is that they exist as discrete molecules. To understand the bonding and stereochemistry in such systems, especially where the metallic element involved is from the transition series, attention focussed on the energy levels arising from the partly filled d-shell, first of all treating the surrounding ligands as a perturbation, lowering the symmetry from spherical, and later by taking account of molecular orbital formation from metal and ligand basis functions. The first approach, borrowed from the solid state physics of transition metal ions in ionic crystals, going back to Bethe in the 1920s, was called crystal field theory, while the second (which built on the concepts of molecular bonding associated primarily with the name of Mulliken), became known as ligand field theory.

However, the very success of these two approaches in rationalising structural features such as stereochemical preferences, and physical properties such as paramagnetic susceptibilities and electronic spectra, served to mask one outstanding weakness. They treated the complexes as if they were totally isolated from one another, without any interaction between them, although most compounds are prepared and studied as solids. Admittedly, the interactions between neighbouring units in the crystals of such compounds are often not large: to use a famous example from the 1960s, the colour of $NiSO_4{\cdot}7H_2O$ as a crystal is almost the same as when it is dissolved in water, because both solid and solution contains $[Ni(H_2O)_6]^{2+}$. Yet there were already many well-known examples to the contrary, some of which were surveyed in the reference just cited. The most spectacular involved electron transfer from one metal ion to another, especially where the metals were in the form of two oxidation states of the same element, so-called mixed-valence compounds (see Chapter 11). On the whole, effects of magnetic exchange interactions between neighbouring metal ions in a crystal cause less spectacular effects on the optical spectra, though it is interesting to note that by the early 1960s Ferguson and McClure were studying the intensification of spin-forbidden ligand field transitions by antiferromagnetic exchange between metal ions in fluoride and oxide host crystals. Many of these studies were driven by a need to understand and master the energy transfer processes in the (at that time) new technology of ruby and related solid state lasers, and scarcely ruffled the surface of the conventional coordination and organometallic chemistry of the day.

Magnetism: From oxides to molecules

Both experimental and theoretical developments in cooperative magnetism of insulating solids (which would later gain significance in molecular-based magnets) were also being driven by technological imperatives. The coming of microwave devices required new magnetic materials that were not metals, i.e. which had only discrete energy states above the ground state, and the many families of ferro- and ferri-magnetic perovskites, spinels and magneto-plumbites were the result. Along with these came the theoretical advances that were to underpin later developments in molecular-based magnetism, such as Anderson's model, based on the Heisenberg–Dirac–Van Vleck approach, systematised into the symmetry rules of Kanamori and Goodenough. It was the latter, above all, that brought cooperative magnetism into the realm of solid state chemistry by setting out clear recipes for predicting (at least in principle) how the sign of the magnetic exchange interaction, and at least qualitatively its magnitude, depends on the number of electrons occupying each orbital on a metal ion and the angle M-X-M′ subtended by two neighbouring metals (M, M′) and the anion X bridging them. The link to molecular-based magnetism becomes clear at once when it is realised that X can be either mononuclear (O^{2-}), bi- or tri-nuclear (CN^-, NCS^-, N_3^-), or much more extended ($^-O_2C{\cdot}C_6H_4{\cdot}CO_2^-$), etc., etc.).

Before advancing further towards the aim of placing Olivier Kahn's achievements in their historical and scientific context, a few remarks are in order about the semantics of the word 'magnet', which carries rather different overtones depending on whether it is used by a chemist or a physicist. To a physicist, a magnet is a macroscopic object containing a near infinite number of atomic moments whose orientations are correlated over a length scale comparable to that of the object itself, i.e. effectively an infinite number of crystallographic unit cells. The object therefore shows a spontaneous (i.e. zero field) magnetisation below a definite phase transition temperature. Chemists, on the other hand, often speak of magnetic ions or molecules, when they really mean 'paramagnetic'. The distinction between the two, and the misunderstandings that have arisen from not taking the difference into account, form an important aspect of the collision between the two scientific cultures of chemistry and physics engendered by Olivier Kahn's work.

On the other hand, the demarcation between infinite and finite spin arrays was already present in the physics world before chemists started making new magnetic objects, through the phenomenon of low-dimensionality. To a physicist, 'infinite' means just that—in all three dimensions. But plenty of inorganic and metal-organic solids exist in the form of layers or chains of strongly interacting atoms or molecules, with only very weak interactions between them. Thus in principle the possibility exists of a crystal with atomic moments correlated over an infinite range in one or two dimensions, but with no three-dimensional order.

Starting from the principles of structural inorganic chemistry, in the 1970s a broad correlation of structure and properties in low-dimensional solids emerged, based on the presence of single or mixed valency, and of ligands bridging the neighbouring metal centres in the lattice. In the same period, low-dimensional magnetism began to interest physicists, especially from the viewpoint of statistical thermodynamics of phase transitions and critical exponents of the various order parameters. To validate their models they measured properties of various prototypes taken from the rich library of inorganic and metal-organic solids, a development that brought them into mutually fruitful contact with synthetic chemists. Examples from this period are the hexagonal perovskite halides with organic cations, $[N(CH_3)_4]MX_3$ and the layer perovskite halides with n-alkylammonium cations, $(RNH_3)_2MX_4$. The latter, in particular, furnished the first series of solids containing molecular units to be recognised by physicists as magnets, in the sense that the layers of ferromagnetically correlated spins were in turn coupled ferromagnetically to give bulk ferromagnets. The M = Cu compounds have sharp phase transitions to long-range order 6–10 K, while the M = Cr ions, having $S = 2$ rather than $S = 1/2$, have Curie temperatures between 35 and 50 K, depending on X and the organic cation.

All of which begs the question as to what really constitutes a molecular, or molecule-based magnet. Here the above examples of chain and layer perovskites provide an interesting contrast. In the MX_3^- chains in the hexagonal perovskites the M-X bond lengths are equal, so the chain is like an infinite polymer. In the layer perovskite halides of Cu and Cr, in contrast, a strong static Jahn–Teller distortion around the metal ions produces an anti-ferrodistortive ordering of the X in the basal plane so that, to the eye of a coordination chemist, the MX_4^{2-} layer appears to consist of well-defined MX_4^{2-} planar molecular ions with neighbouring molecular planes all orthogonal. This cooperative Jahn–Teller ordering results in a lattice in which neighbouring M orbitals carrying the unpaired electrons responsible for the magnetic moment are constrained to be orthogonal. That is the classic Kanamori–Goodenough rule for ferromagnetic exchange.

It was one of Olivier Kahn's early successes to engineer a combination of d^1 ion (VO^{2+}) and a d^9 one (Cu^{2+}) into a dimeric unit using the specific ligating capabilities of a Schiff base anion in such a way that xy and $x^2 - y^2$ orbitals on the two metals were also orthogonal, inducing a very large ferromagnetic exchange interaction between them. Nevertheless, such a dimer is not a magnet. It is also an inconvenient fact that more exchange pathways between localised atomic moments through bridging ligands lead to antiferromagnetic exchange than ferromagnetic. To engineer infinite lattices

Fig. 34. Molecular-based magnet in the form of a 'necklace'.

containing bridging ligands, Kahn therefore turned to ferrimagnetism, using metal ions (Cu^{2+} and Mn^{2+}) having the largest difference in moment possible for 3d ions. Further ingenious modification of ligand substituents by adding –OH groups led to ferrimagnetic chains, interacting to give a crystal with 3D ferromagnetic order. Unusual lattice topologies also emerged from bimetallic combinations of the same elements, including interlocked orthogonal hexagonal networks, which (as Kahn said more than once with a typical rhetorical flourish), reminded him of a necklace that he had given to his wife (Fig. 34).

A scientific legacy

Many more examples could be cited of Kahn's focussed imagination in devising new solid state coordination compounds with unusual structures and properties. High coercivity magnets, for example, based on strikingly elaborate structures containing $Mo(CN)_7^{4-}$, and interlocked networks, or spin-crossover compounds tailored beautifully to exhibit sharp hysteretic transitions straddling room temperature and therefore, as a result of the colour different between the spin states, able to act as thermochromics. But, above all, what could be claimed as Kahn's scientific legacy is the growth and increasing maturity of the whole field of molecule-based magnets.

With the enormous number of conventional magnetic materials available to technology as the outcome of a century's progress, it is legitimate to ask what fundamentally new features the molecular solid state introduces into the field. In my view there are three. First, and by no means least, is the issue of processing. Until the coming of molecule-based magnets all known magnetic materials had to be synthesised at high temperatures by ceramic or metallurgical methods. Now there is a serious prospect of soluble magnets, or at least materials that can be deposited in thin films by dip- or spin-coating. Second, in contrast to continuous lattice magnetic materials, the vast majority of which are either metals or small band gap semi-conductors, molecular-based magnets are insulators whose lowest energy excited states are discrete and localised. This opens up the possibility of spectacular photomagnetic and magneto-optical effects. Combining the first two of these attributes, one of my own graduate students 20 years ago said of such a compound he had made that it was a green ferromagnetic soap! The third attribute of molecular-based magnets, potentially the most fruitful in the long term, is the opportunity it gives to harness the whole variety of synthetic molecular chemistry, organic as well as inorganic, to devise lattices capable of supporting properties (or combinations of properties) not attainable with simpler continuous lattice solids. Properties based on a combination of magnetic order and structural chirality come to mind, or new lattice topologies, of which the Kagome (whose critical behaviour is intensely interesting to statistical thermodynamics) is only one simple example.

Seen from this point of view, molecular-based magnetism falls into place as one of many fruitful outcomes of the evolution from molecular to supramolecular chemistry. It is a flourishing department in d- and f-block coordination chemistry, and holds the promise of eventual technological outcomes in data storage and display. Considerable credit for that state of affairs rests firmly with Olivier Kahn, whose combination of talents gave the field the impetus to develop so strongly. Was Cleopatra's nose the right length? In this case the answer must be an unequivocal 'yes'!

chapter 9

Fred Dainton: Scientist and Public Servant

Fred Dainton was one of those still comparatively rare animals, a member of the higher mandarinate who was not only educated as a scientist, but had a distinguished academic and research career before immersing himself full-time in administrative affairs. His background in the sciences equipped him with an incisively analytical approach to problem solving, but he was also a man of extremely wide learning and intellectual sympathy, and (it should be said at the outset) remarkably good company, provided one was not advancing a proposition on less than secure logical foundations. Then he could be quietly and courteously devastating.

Fred Dainton came up the hard way, the ninth child of poor parents in Sheffield: his father was a stonemason who had not had schooling enough to read and write though, perhaps because of that, was a fervent believer in the value of education. Fred won a scholarship to the Central Secondary School, Sheffield, where his long journey to academic distinction began. At St. John's College, Oxford, where he was an Exhibitioner, he was one of the many distinguished pupils of H. W. (Tommy) Thompson reading chemistry, but after getting a First in 1937, somewhat unusually he did not remain at Oxford to begin his research career but moved to Cambridge to work on the reactions of simple gases with R. G. W. Norrish, who later won the Nobel Prize with George (now Lord) Porter for work in the same field.

It is possible that his relations with the flamboyant 'Tommy' may not have been of the best. Be that as it may, the first part of his career as an academic teacher began in Cambridge and he was not to return to Oxford again till much later; though return he did, and in fact made his last (certainly it should not be said retirement) home in the city.

Shortly after he started his first teaching appointment, war broke out and Dainton was called to secret research on the effects of radiation, a topic that was to recur later as he took up an independent line of research after the war. A Fellowship at St. Catherine's Cambridge followed, held until 1950 with a University Lectureship. It was during this war period in Cambridge that he met and married Barbara Wright who had read zoology at Newnham and was subsequently to hold a Fellowship at St. Hilda's in Oxford. To the end they formed an outstanding partnership, Barbara's support taking many forms, not least, when Fred was separated from secretarial backup after retirement, in keeping track of his extensive correspondence and preparing papers for House of Lords Select Committees. Many members of St. John's College will remember their regular attendance together in Chapel on Sunday evenings when both in their eighties, followed by dinner in Hall. Indeed, the last words that Fred spoke to me, only a few weeks before he died, was to admonish me for not appearing more often at High Table on a Sunday evening, and suggesting that Frances and I should make up a foursome with him and Barbara before too long.

Fred Dainton's administrative skills were quickly spotted and in 1980 he was invited to take the Chair of Physical Chemistry in Leeds, where he founded the Cookridge High Energy Radiation Laboratory, attached to the local hospital. His research into the chemical effects of radiation flourished at Leeds and seminal work on the reactions of solvated electrons earned him election to the Royal Society at the comparatively early age (for a chemist) of 43. Subsequently he also received the Society's highest award for chemistry, the Davy Medal. But University administration beckoned and he left Leeds to become Vice-Chancellor at Nottingham. This was the time of student militancy, when he had his share of quelling demonstrations, though his preferred method of rational discourse, leavened with irony, cannot have endeared him much to the wilder of his flock. In 1970 the opportunity to follow his revered mentor in reaction rate theory, Hinshelwood, as well, perhaps, as a certain exasperation with student politics, brought him back to Oxford as the Dr. Lees Professor of Physical Chemistry. It was in this guise that I first encountered his formidable presence, on an obscure body called the Chemistry Sub-faculty Steering Committee. At the first meeting after his arrival he made little contribution to the discussion but the second was different: at the earliest opportunity he began quietly 'I have always made it a rule not to intervene in the debate the first time I go to a new committee, because I like to understand what the important issues are first'. Then he proceeded to demolish the proposals we were considering with a sequence of quietly courteous but devastatingly accurate blows. One had the feeling, though, that in this small pool he was only boxing with one hand tied behind his back.

Bigger issues commanded his attention when he became Chairman of the University Grants Committee three years later, and he left Oxford for the second time. Life under the then Secretary of State, Margaret Thatcher, cannot have been easy for a man of Fred's sympathies, though their shared love of forthright debate must have led to some fascinating passages of arms. A last big job was the Chairmanship of the Board of the British Library, which arose out of his earlier work on a group set up by the previous Labour government to examine the needs of all the national libraries. He it was who presided over the planning of the institution that finally opened its doors to readers at St. Pancras, and insisted that the new building should take the fullest account of advances in information storage and retrieval.

The last 11 years of his life saw the stonemason's son a respected member of the House of Lords, frequently called on to take part in inquiries by the Select Committee on Science and Technology, and Chancellor of the University in the city of his birth. A flavour of Fred Dainton's pithy Yorkshire style can be glimpsed in the following anecdote told against himself (with a quizzical lift of the eyebrow): walking high in the lonely Yorkshire Moors one fine day, Fred met an old countryman coming towards him. Fred greeted him courteously, remarking what a splendid day it was; the old man paused in thoughtful silence for a moment, and delivered himself of a put down that even Fred would have found hard to counter: 'Aye', he said, 'now't wrong with t'day, its folks you meet'. Among many whose paths to distinction have passed through St. John's College over the last half century or so, few achieved more, or were better folks to meet than Fred Dainton.

part 4

MOLECULES, SOLIDS AND PROPERTIES

The popular press finds it impossible to write about any progress in science without using the word 'breakthrough' in the first sentence. Yet is this really a proper view of how science goes forward? Even if we grant that newspapers thrive on sensation and find 'the trivial round, the common task' insufficiently dramatic to sell many copies, there is still a sense in which understanding advances by fits and starts. Knowledge, in the sense of observations accumulated, progresses in a more even flow; it is what we make of those observations that is uniquely subject to the vagaries of the human imagination—and all the more fascinating for that.

Naturally, in this (as in so many aspects of the human intellect) the philosophers got there first and Thomas Kuhn, writing on 'The Nature of Scientific Revolutions' coined the phrase 'paradigm shift' that has passed into common discourse for almost any situation where a new way of looking at an old problem needs to be aphorised. But collecting facts and then explaining them (the Baconian model, if you like) is a very imperfect way to describe what actually goes on in science. Where we choose to look in the jungle of an untidy world is conditioned very much by prior expectations about what we hope to find. Recall the story of the man looking for a lost key under a lamp in a dark street. On being asked whether that was where he lost it, he replied that he didn't know, but that was the only place where it was light enough to look.

For some reason, when the philosophers of science such as Kuhn seek instances of paradigm shifts, it is nearly always to what we call nowadays 'big science' that they look. The Copernican revolution, the dawn of relativity and the replacement of the steady state theory of cosmology by the 'big bang' are common examples. But the less apocalyptic (but more immediately valuable) sciences of bulk matter such as condensed matter physics and solid-state chemistry have their own powerful lessons to teach us when we want to see how entrenched mindsets can be overturned and a subject sent off in a wholly new direction. The three examples from basic chemistry and physics of solids that follow are (like the places and the people described in earlier sections) ones which I have lived through myself and, hopefully, contributed towards shaping. They show, too, how paradigm shifts can be impelled by very different kinds of impetus.

Until a decade ago, the word 'magnet' would have automatically summoned up an image that would have been familiar to Faraday, a shiny metal bar picking up iron-filings. Now chemists are making magnets in their flasks and beakers that are certainly not metals, and may not even contain any metallic elements. In the case of mixed-valence, it was a matter of a large number of disparate observations about colour and chemical constitution that had been lying around, in some cases for a century or more. A single, simple unifying hypothesis was enough to bring this particular terrain of the chemical jungle into instant order and cultivation. On the other hand, superconductivity (for which mixed-valence is often an important adjunct) has always seemed a bizarre and improbable phenomenon. Its discovery at much higher temperatures than hitherto thought possible, and in a class of mixed-valence ceramics where it had not previously been anticipated, however, was a genuine bolt from the blue: a new fact that changed perceptions overnight (or, at least worldwide, in about ten days).

My final examples in this section are closer to technology than basic science but also illustrate how illuminating it can be to approach an old subject from a new angle. The prefix 'nano-', applied to just about anything, arrived with a bang in the newspapers only a few years ago, though chemists especially have known perfectly well for most of the twentieth century that this length scale (10^{-9} metre) was precisely the one that their own science had operated on all that time. Nevertheless the new label has created a new focus for technology, especially where the storage and processing of information is concerned. How molecule-based information processing will play out, nobody knows yet but gazing into crystals can be fun for chemists as well as for astrologers.

chapter 10

Magnets from Molecules

The Pre-History

Does the history of science matter to the progress of science? Without going so far as to agree with Henry Ford's famous dictum that 'history is bunk', there is undoubtedly a sense in which the answer must be 'no'. If the point of science is to uncover and then explain the facts of the natural world, and the facts themselves are objective, it is not going to matter much who made a particular observation, or who was the first to expound a particular model. If Einstein had not had the idea of special relativity in 1905, or if Faraday had not discovered the phenomenon of electromagnetic induction in 1831, we can be fairly sure that within a few years somebody else would have done so. The reason that we can be so certain about this lies in the nature of science itself as a cooperative social process. Individual scientists, even those as great as Faraday and Einstein, do not work in isolation from the rest of the scientific community: the sheer volume of Faraday's scientific correspondence is ample testimony to this point. With only the rarest of exceptions, and especially with all the modern means of communication, they are fully (although perhaps only subconsciously) aware of the state of knowledge, not only in their own field but also in cognate fields.

If such a model of science as a complex interacting social system is accepted, then in trying to understand after the event why a certain development took place when (and in the way) that it did, we do need to look at the history, or at least the chronology, of events. In life, generalizations have to be backed up by examples, just as, in science, theories need to be validated by facts. In this chapter I take one currently fashionable topic, molecular magnets, and look into its antecedents a little more closely. For example, recent review articles on molecule-based magnets make the following statements:

> *Molecular-species based ferromagnetic compounds, although postulated in the 1960s, have been realised only within the last decade.*

> *The first molecular compounds exhibiting a spontaneous magnetisation below a critical temperature were reported during the 1980s.*

Neither statement is correct.

For the sake of clarity we can agree that a magnet is a solid that exhibits spontaneous (i.e. zero-field) magnetization below a critical temperature, and by 'molecule-based' one can construe a solid whose structure consists of recognizable molecular units bound together by ionic, covalent or van der Waals interactions, and usually assembled from solution. What is beyond doubt is that until the 1950s no such materials had been found. In part, at least, that is because nobody had looked, although there is naturally a more profound reason.

Inorganic chemistry and solid-state physics

From World War II into the 1960s, inorganic chemistry had bifurcated: there was an explosion of interest in coordination chemistry, symbolised on the one hand by the inauguration of the International Conferences on Coordination Chemistry in 1949 (which still continue), and on the other by what might be called classical solid-state chemistry, which dealt with continuous lattice oxides, halides, chalcogenides, and so on. Crystals formed by coordination complexes are, of course, molecular but the bulk properties of such solids were rarely, if ever, considered. That was because the dominant (and, within its limitations, very successful) theoretical model used to rationalize the structures and electronic properties of coordination complexes, namely ligand field theory, took as its starting point the d-orbitals of a transition metal ion, more or less perturbed by the ligating atoms. Thus, attention was thrust on to the individual molecular entity, anything beyond the nearest-neighbour ligands being neglected.

Conversely, much of the impetus behind the solid-state chemistry of oxides during the same period came from technology, in particular microwave communications and information storage. Ferrites, garnets and magneto-plumbites were being thoroughly researched as memory devices, modulators and recording media, which in turn required a better understanding of the microscopic magnetic interaction mechanisms. At the phenomenological level of the molecular field model, this had been accomplished by Neel in his 1948 study of ferrimagnetism, but it was not till a few years later that Anderson described a quantum mechanical approach to the interaction between spins localised on neighbouring metal ions mediated by an intervening closed-shell anion such as the oxide ion. However, the language of that model was more opaque than most solid-state chemists could handle, and it was only after a further interval of a few years that Kanamori and Goodenough translated it into a series of symmetry rules.

They defined the sign of the magnetic exchange interaction between two metal ions M and M′ separated by an anion X according to the numbers of unpaired electrons occupying the orbitals containing the unpaired electrons and the angle M-X-M′, with the limiting cases of 90° and 180°. Most tellingly, they noted that if orbitals carrying the unpaired electrons on the two metal ions are orthogonal to one another, the interaction is ferromagnetic; this was in fact an extension of Hund's rule from the 1930s. That put a predictive weapon into the hands of the experimentalists. With the result that an enormous number of complex oxides and sulphides were synthesised and magnetically characterised, as summarised, for example, in Goodenough's magisterial book 'Magnetism and the Chemical Bond.'

Magnets and coordination complexes

However, where were the coordination chemists while all this was going on? The only coordination compound in the 1950s whose unusual magnetic properties were shown conclusively to arise from interaction between metal ions was dimeric copper(II) acetate, the diamagnetism of which at low temperature was quantitatively explained in terms of antiferromagnetism by the physicists Bleaney and Bowers. The same group at the Clarendon Laboratory, Oxford, also uncovered the first case of magnetic exchange interaction between two metal ions not directly connected by bridging ligands when they found that the electron paramagnetic resonance spectrum of the hexachloro-iridate(IV) anion doped in the diamagnetic host lattice of potassium hexachloro-platinate(IV) varied with doping level, because pairs of anions formed at higher concentrations. The pure compound potassium hexachloro-iridate(IV) becomes antiferromagnetically ordered at very low temperature. Although present in an ionic lattice, the hexachloro-iridate(IV) anion can hardly be called anything but molecular.

Nevertheless, those examples are antiferromagnets; what of ferromagnets? As the quotation at the beginning of this chapter avers, it was indeed in the 1960s that a potential mechanism for ferromagnetic exchange between localised moments in an insulating molecular crystal was put forward. Nevertheless, it was a decade before then that the first genuine example of bulk ferromagnetism in molecule-based crystals came to light at Bell Laboratories. It is true that the paper reporting this seminal result is only half a page long, but the abstract is unequivocal; I quote, in full: 'Certain of the complex cyanides of elements of the 3d transition group appear to be ferromagnetic at very low temperatures'. As the satirical magazine Private Eye might say, '. . .er, that's it!' The evidence adduced was twofold: a maximum in the susceptibility and a remanence. The compounds in question were transition-metal cyanide complexes of the Prussian Blue type and the physicist authors were suitably reticent about their precise chemical formulae. One is reminded of a similar reticence, some 30 years later, by the authors of the first paper reporting high temperature superconductivity. The role of the theoretical physicist P. W. Anderson, the author of the super-exchange model mentioned above, as a co-author of the brief experimental paper on ferromagnetism in molecule-based solids is certainly worthy of note.

Prussian Blue itself, which has been known for about 300 years (Fig. 35(a)) was first subjected to study by neutron diffraction, the definitive technique for verifying the order of magnetic moments in solids, in 1973 (Fig. 35(b)), and by polarized neutron diffraction only a few years later. Unlike other bimetallic hexacyano-salts in which both metal sites carry unpaired spins, the fact that Prussian Blue itself is a ferromagnet seems rather strange at first sight because the Fe^{II} is low spin and therefore diamagnetic. Moreover, in the simple cubic lattice of alternating Fe^{II} and Fe^{III}, bridged by CN groups, nearest-neighbour Fe^{II} with unpaired electrons are

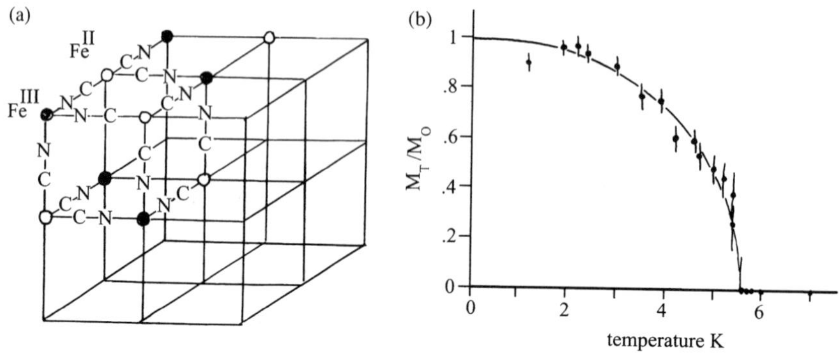

Fig. 35. Prussian Blue: (a) crystal structure; (b) magnetization *vs*. temperature by neutron diffraction.

separated by no fewer than five closed-shell atoms (Fe^{III}-NC-Fe^{II}-CN-Fe^{III}) at a distance of no less than 10.2 A. Admittedly the Curie temperature is low (5.5 K), but the surprise is that it orders at all.

Prussian Blue is, of course, the grand-daddy of all mixed-valence compounds, a subject that had attracted my own attention in the 1960s, and in 1975, with a graduate student Bryan Mayoh, I published an article with the title 'Contribution of mixed valence to the ferromagnetism of Prussian Blue'. In it we showed quantitatively how an admixture of the low-lying inter-valence Fe^{II}-Fe^{III} excited state with the ground state stabilizes a parallel rather than an anti-parallel arrangement of the Fe^{III} spins. Much more recently, a systematic synthesis of other mixed-valence cyanides with Prussian Blue structures has driven the Curie temperatures of this compound type even higher, so that they are now well above room temperature.

Argument of a somewhat semantic nature has turned on what really is a 'molecular' magnet; is it a solid constructed from molecular building blocks (what the late Olivier Kahn—Chapter 8—called 'bricks') or should one restrict the phrase to solids in which there are no direct bridging groups between the magnetic units? Prussian Blue would be excluded by the latter criterion because it consists of an infinite three-dimensional network of coordinate links Fe^{II}-C and Fe^{III}-N. It seems clear that the first strictly molecular solid definitively verified as sustaining ferromagnetic order is the rather unlikely compound diethyl-dithiocarbamato-Fe^{III}CI. How it came to light is itself a fascinating instance of serendipity.

The first molecular ferromagnet

During the 1960s there was much interest in making coordination complexes that would mimic the active sites in metallo-enzymes, among them the FeS proteins such as ferredoxin. Fe complexes of dithio-carbamates (Fig. 36(b)) made for this purpose proved to have some quite strange and unlooked for properties, such as intermediate spin states of Fe^{II} between $S = 0$ (low spin) and $S = 2$ (high spin) and also temperature-dependent spin crossover from one spin state to the other. The latter has burgeoned into a subject in its own right, with implications for data storage devices. Over the same period, Mössbauer spectroscopy was coming into its own as a tool for identifying electronic ground states, and thus it was that one (and only one) of the dialkyl-dithiocarbamates of Fe^{III} proved to exhibit magnetic hyperfine splitting at low temperature, indicating that there was an internal magnetic field within the lattice (Fig. 36(a)). The zero-field a.c. susceptibility identified T_c as 2.5 K. The crystal structure is uncompromisingly molecular, with no intermolecular contacts shorter than van der Waals radii. Subsequent very detailed measurements of single crystal susceptibility fully confirmed the ferromagnetic order and the easy axis of magnetization. Curiously, changing the alkyl group and the halide failed to produce any more magnets of this kind.

Not long afterwards, another purely molecular solid with spontaneous magnetization at low temperature came to light in the form of Mn phthalocyanine. The phthalocyanines containing a divalent transition metal form an isomorphous series in which the planar molecules form stacks, arrayed in a herringbone fashion so that an N atom on one molecule lies above the metal atom at the centre of the next (Fig. 37). Except for the Mn compounds they were antiferromagnets at low temperature but the latter proved to have an appreciable saturation magnetization, indicating that the moments were substantially canted, giving rise to weak ferromagnetism.

Low-dimensional molecule-based magnets

During the 1970s further examples of ferromagnetic exchange between simple coordination complexes were synthesized and, indeed, predicted by following the Kanamori–Goodenough recipe of maintaining orthogonality

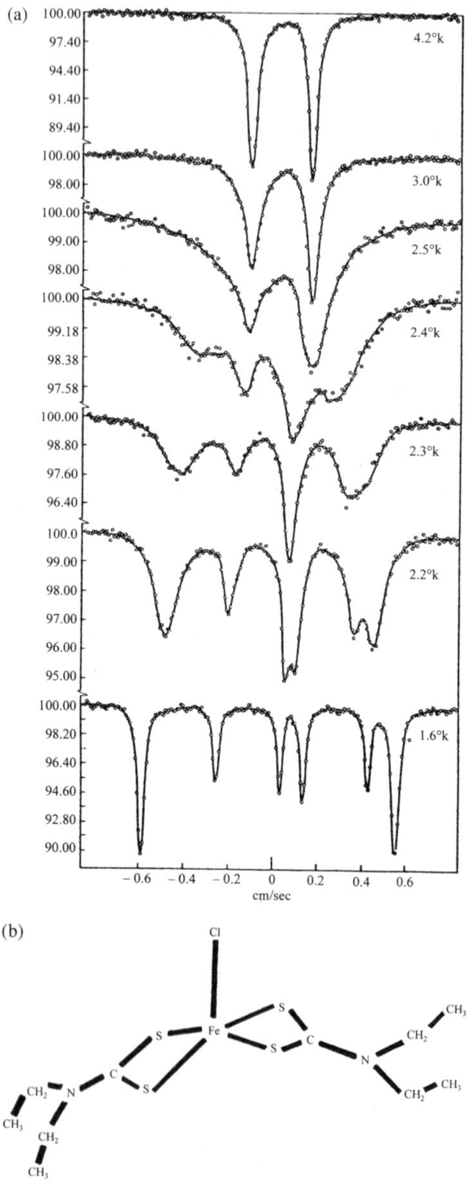

Fig. 36. *Bis*-diethyldlthiocarbamato-FeCl. (a) magnetic hyperfine splitting of the Moessbauer spectrum showing the internal magnetic field; (b) the molecular structure.

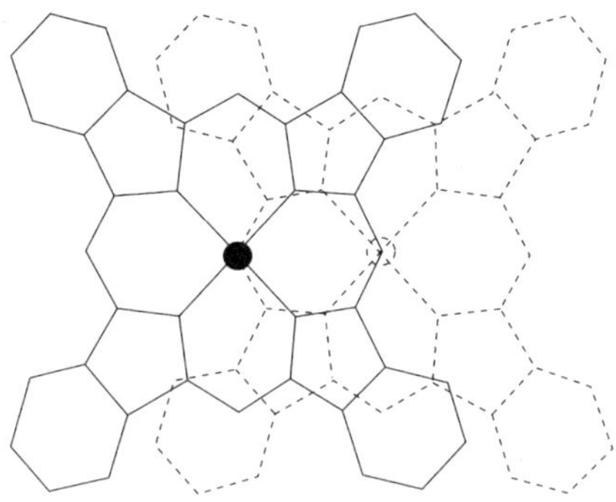

Fig. 37. Molecular packing in the crystals of metal phthalocyanines.

between magnetic orbitals on neighbouring metal centres. The first cases of this type were the so-called layer perovskite halide salts of Cu^{II}, whose critical behaviour was closely studied by the Leiden group as test cases for the statistical thermodynamics of the $S = 1/2$ square-planar Heisenberg ferromagnetic Hamiltonian. Meanwhile in Oxford, we realised that the first strictly molecular solid definitively verified as sustaining ferromagnetic order is the orbital orthogonality induced by cooperative Jahn–Teller distortion of the coordination environment, and T_c values up to 50 K were found. In the Cr series there was the added bonus of extraordinary optical properties, whereby the ligand field transitions changed intensity by several orders of magnitude from the paramagnetic to ferromagnetic state, thus causing a colour change visible to the naked eye. These compounds, like the cyanides already described, and the bimetallic oxamido-ferrimagnets synthesized by Kahn and his colleagues in the 1980s, somewhat beg the question as to what is 'molecular' because the planar MX_4 units are organised into layers with neighbouring planes orthogonal via weak intermolecular M...X contacts (Fig. 38). However, they are all prepared from solutions of their molecular constituents.

A mature discipline emerges

Some years later, the 1980s saw a further step forward with the synthesis of the organo-metallic charge transfer complex decamethylferrocinium-tetracyano-ethylene, which becomes fully ferromagnetic at 4.8 K and is

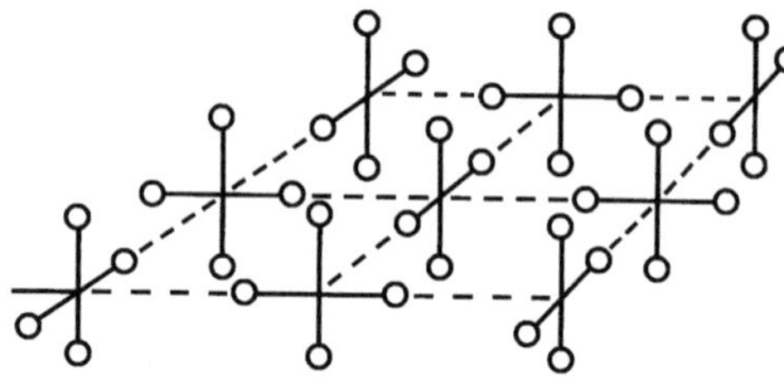

Fig. 38. Schematic crystal structure of alkylammonium tetrachlorochromates(II).

entirely molecular, with the moment on the tetracyano-ethylene anion mediating magnetic exchange between the Fe moments. Early in the following decade came an even more exciting departure in the field, the first molecule-based ferromagnet that is purely organic, in other words in which the unpaired spins are p rather than d or f. Thus we come to the present day, when hundreds of coordination, organometallic and even organic chemists throughout the world are engaged in the search for new molecule-based magnets with higher T_c values, other properties characteristic of the molecular solid state such as mesomorphism and chirality, and with the first technological applications being close. The European Science Foundation has sponsored a Programme on Molecular Magnets, the European Commission funds three networks on different aspects of the field, and a biennial international conference attracts hundreds of participants. This brief account of the antecedents leading up to the present burst of activity in this interdisciplinary field will serve to highlight some of the influences that shaped it.

The Chemistry of Magnets

Magnetism and chemistry have had an up and down history together. From very early times, chemists thought that useful information about the bonds between atoms in molecules and solids could be derived from measurements of magnetic susceptibility. Faraday spent much time hanging samples of extremely diverse materials between the pole pieces of an electromagnet that he himself had made to detect differences of behavior between, as he put it, 'a piece of wood, or beef, or apple obedient to or repelled by a magnet.'

More recently, and more quantitatively, Pauling's seminal work of the 1930s on 'The Nature of the Chemical Bond' contained a lengthy section on 'The Magnetic Criterion for Bond Type', and in the 1950s, the advent of ligand field theory made the measurement of bulk magnetic susceptibility respectable as a way of learning about metal-ligand covalency and the symmetry of coordination around transition metal ions. However, despite the fact that bulk susceptibility measures just that—the reaction of the whole sample to an applied field—until recently, only a small minority of chemists have interested themselves in the effects of interaction between the neighboring metal ions in an extended lattice.

Meanwhile, the world of 'real' magnets, dominated by physics and materials science, goes on, as it has since the time of Faraday, in its preoccupation with metals, objects that are opaque, shiny, and go clang when they hit the floor. These materials do not have much in common with the kinds of compounds that chemists make in their flasks and beakers, you may think. Furthermore the information-storage industry (which embraces audio-tapes as well as more sophisticated hard disks) relies heavily on transition metal oxides, whether simple binary ones like γ-Fe_2O_3 and CrO_2 or complex solid solutions based on the garnet or ferrite structures. Still, all the while, a small dedicated band of chemists has been trying to synthesise substances that have a bulk spontaneous magnetisation.

For example the synthesis of a material, formed from perpendicular interlocking graphite-like networks of spin centres, that exhibits bulk magnetisation below 22.5 K, is an excellent example of the genre (Fig. 34). Given that so many conventional magnetic materials are available with Curie temperatures (T_C's) far above room temperature, not to mention high coercive and remanent fields, it is worth asking why so much effort is being devoted to the synthesis of substances with Curie temperatures of only slightly above the boiling point of liquid helium.

One reason, by no means to be ignored, is the sheer novelty of this approach to cooperative magnetism. Making new molecular architectures, tailored to have a specific property, is very much a task for chemists. Another important feature characterising such non-conducting magnets is their transparency, certainly not something normally associated with spontaneous magnetisation. For example, magnets whose optical spectra consist of discrete absorption bands may change their transparency or birefringence quite dramatically when long-range spin correlations build up as a function of temperature or applied field. Should transparent materials be synthesised with spontaneous magnetisation close to 300 K, applications in data storage and telecommunications would surely follow.

There are several other features of potential practical import that distinguish magnetic materials based on molecules from their analogues consisting of continuous ionic or metallic lattices. Synthetic methods are, of course, quite different, based on the traditional methods espoused by coordination or organometallic chemists, such as crystallisation from solution close to room temperature. Consequently, thin films might be deposited with methods, such as solvent evaporation or spin-coating, that are familiar from polymer science.

Could magnetisation be associated with other properties found exclusively in molecular assemblies such as mesomorphism? For the time being, we do not know. A variant on molecular-based magnets is to enrobe clusters of inorganic material with an organic coating, on the way to mimicking biological objects like ferritin or the small particles of magnetic oxide used by bacteria (or even pigeons) as a means of navigation. Figure 39 shows such a system containing 19 exchange-coupled Fe^{III} centers.

If all of the above are taken as good and sufficient reasons to try to make molecular-based magnets, how should the job be tackled? The first point is that to make magnetic materials that are not metallic conductors, one can concentrate on designing lattices in which the magnetic exchange interactions that order the localised moments are principally between nearest neighbours. Sometimes it may be necessary to think about next-nearest neighbours and even, quite exceptionally, third-nearest neighbours. Long-range exchange, of the RKKY type for instance, which involves conduction electrons, does not have to be considered. From that point, the options are easily defined: clearly, one can try to engineer near-neighbour ferromagnetic exchange, but exploiting ferromagnetism is also an attractive possibility. In the latter, the exchange interaction between dissimilar ions is antiferromagnetic, but because they have different moments, there is a net resultant moment. A third option is to make lattices containing antiferromagnetic near-neighbour interactions of sufficiently low symmetry that the moments are not exactly anti-parallel but make a small angle to one another, called

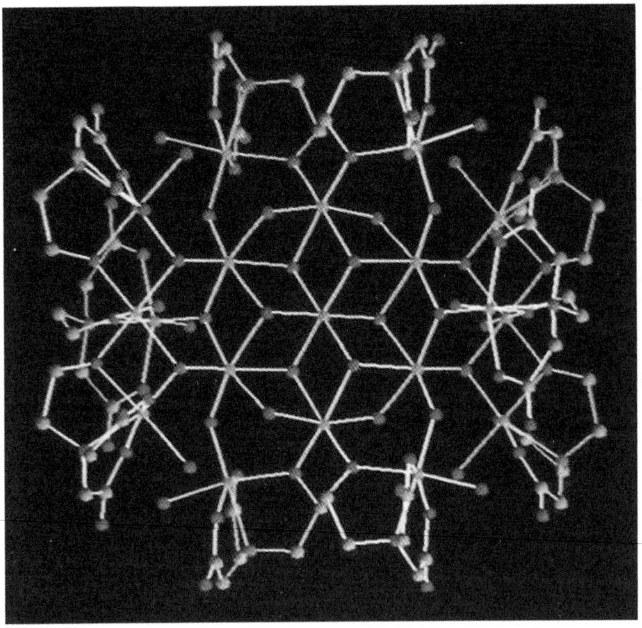

Fig. 39. One approach to the synthesis of molecular magnets is to create hydroxo(oxo)polyiron complexes, as shown here.

canting. Then there remains a small, uncompensated component giving what is called weak ferromagnetism.

All these strategies have been tried, and over the last 10 to 15 years, all have led to insulating magnetic compounds. The rules relating the symmetry and occupation of orbitals on neighbouring centres to the sign of the exchange interaction have been known for nearly 40 years under the names of Kanamori and Goodenough, so that (for example) combinations of octahedrally coordinated Cr^{III} and Ni^{II} linked through a bridging ligand are invariably ferromagnetically coupled.

Simpler predecessors of the elaborate networks such as in the compound in Fig. 39 were built from metal complexes, with ambidentate ligands, that form bimetallic chain polymers. The recipe is straightforward: make a complex of metal A with a ligand that is capable of binding a second metal B on its back side. The polymers have to be built into three-dimensional arrays in such a way that interactions between the nearest ions on adjacent chains allow the spins to add and not subtract, a tricky problem to which much experimental imagination has been devoted. Transmission of the interaction through bridging organic molecules has also been explored by the construction of discrete dimers, blocking off further polymerisation with capping ligands. Typical bridging ligands are oxalate, dithio-oxalate, and oxamate. Very large intra-molecular ferromagnetic exchange arises in dimers when a transition metal ion with a single d electron is combined with one containing a single hole in an otherwise filled d shell, specifically VO^{2+} and Cu^{2+}. That is because the single electron centred on the V is in an orbital of xy type, whereas the hole on the Cu is in an $x^2 - y^2$ orbital.

This illustrates an important general recipe for ferromagnetic exchange, namely that the orbitals bearing the moments must be orthogonal. Unfortunately, except in rather specialised cases, it has not proved very easy to make networks obeying this principle, though one method that I used some 25 years ago exploits the Jahn–Teller effect as a way of spatially ordering the moments. That is why ferrimagnetism, based on two different antiferromagnetically coupled sub-lattices, is more commonly used. The biggest possible saturation magnetisation occurs when the two metal ions have as large a difference as possible in spin quantum number. Among transition metal ions, that condition leads one to combine Cu^{2+} ($S = 1/2$) with Mn^{2+} ($S = 5/2$). Pursuing the same line of thought further, several groups considered combinations of 4f and 3d ions, the former being often Gd^{3+} ($S = 7/2$). The Achilles heel of that approach, however, is the small exchange constant resulting from small overlap between well-shielded f and 3d orbitals.

Although they are prepared from molecular precursors, the magnets discussed so far are not, strictly speaking, molecular themselves but are really polymeric. Examples of spontaneous magnetisation among truly molecular solids—that is, ones in which the intermolecular contacts are all at van der Waals distances—are extremely few and far between. To be precise, just one compound of this type has been uncovered and thoroughly investigated: $Fe[C_5(CH_3)_5]_2(TCNE)$ (tetracyano-ethylene), made by Miller and Epstein in the late 1980s.

This material has a chain structure based on alternating organometallic cations and TCNE anions that come together plane-to-plane. Its T_c is 4.8 K, and it has a relatively large coercive field of 1000 G. The mechanism of spin alignment is almost certainly different from that in coordination polymers, in which super-exchange provides the answer. The most plausible explanation for the ferromagnetic sign was first proposed back in the 1960s by McConnell. He suggested that if one of two alternating molecular ions forming a chain had two molecular orbitals of similar energy, the transfer of an electron from one ion to the other would stabilise the arrangement with the maximum number of unpaired electrons. The conditions for forming such an arrangement must he extraordinarily demanding because Miller and his colleagues synthesised some dozens of salts closely related to $Fe[C_5(CH_3)_5]_2(TCNE)$ without finding any others that were ferromagnets.

Apart from transition metal ions, another source of degenerate orbitals to test the McConnell mechanism would be a pπ system of an aromatic organic molecule with high symmetry. But unpaired p-electrons in organic chemistry usually go under the name of free-radicals because they are chemically very reactive. Therefore, the first thing to do when trying to make purely organic magnets is to shield the unpaired electrons from the outside world with bulky substituents. In that way, high-spin molecules have been prepared, although this is only the start: The molecules must interact ferromagnetically with each other to make a bulk material. In fact, the first bulk ferromagnet containing only elements of the first short row of the periodic table was only synthesised a decade ago (see later), although a small family of related compounds is being built up quite rapidly. The molecules in question are nitronyl-nitroxides, each of which contains two NO groups connected by a single =CH–, carrying a single unpaired electron. Variation of the substituents makes possible some crystal structure engineering leading to correlations between magnetic properties and molecular packing. The bad news is that the Curie temperatures are extremely low. In fact the current record is only 1.48 K.

There is a long way to go before we shall see molecular-based magnets turning up in audio tapes or storage disks. On the other hand, the efforts so far have turned up a lot of new chemistry and brought synthetic chemists into close contact with physics and materials science. In Europe, the latter is being helped through funding by the European Commission. Momentum in that part of the world is being matched in Japan and the United States, showing that chemists are becoming enthusiastic about magnetism again.

Magnets Without Metals

As bars, horseshoes or other more exotic shapes, magnets have been a familiar part of our lives since the days of lodestone. They turn up in nearly every domestic electrical device, and give us the means of recording information on tapes and disks. But what are magnets made of? After lodestone, soft iron in the early days: the Royal Institution has some beautiful examples in its museum which were used by Faraday. In the past 10 years, advances in metallurgy and solid-state physics have given us new magnetic materials such as the rare-earth-cobalt alloys and neodymium iron boride capable of producing very high fields from small volumes (in the words of the trade, they have a high magnetic coercivity). All magnets lose their magnetism above some critical temperature called T_c (the 'c' stands for Curie). 'Permanent' magnets have T_c's far above 300 K, so they are fully magnetised at room temperature. Why, then, should a report that a new magnet has been found with T_c of only 1.48 K have been of any interest? As always in materials science, the answer lies in the elemental composition and crystal structure of the new phase.

The crystal structures of the magnets that we are most familiar with consist of continuous lattices of atoms—they are not built from molecules or finite clusters of atoms. Furthermore, such magnets are metals, conductors of electricity. It is also the case that most non-metallic chemical compounds containing the same elements, such as iron or neodymium, have unpaired electron spins that order antiferromagnetically. That is, the spins on neighbouring metal ions line up in antiparallel fashion so that there is no net resultant magnetic moment.

Chemists, being of a cussed disposition, had been trying for some years to make molecular-based compounds that were electrically non-conducting but at the same time magnetic. The reason for such an effort is, of course, not just to overturn the prevailing generalisations, but also to create magnets that, unlike metals, are transparent to visible light. Chemists made some progress in the 1970s in synthesising materials that changed colour when traversing T_c, but more recent work has concentrated on transition metal or rare-earth coordination complexes as the building blocks (or bricks, as Olivier Kahn—one of the leading exponents of this approach—has called them). Ingenious molecular design ensures that the unpaired electrons on neighbouring metal ions in the crystal occupy orbitals orthogonal to one another, the quantum mechanically based recipe for ferromagnetism. Compounds of this type have been prepared with T_c as high as 30 K. But why the fuss about a compound with T_c of 1.48 K?

In the earliest days of quantum mechanics, Heisenberg predicted that ferromagnetic long-range order in an infinite lattice would be found only when the lattice contained heavy (that is, metallic) elements. Hence another challenge to the chemist: could one make a molecular ferromagnet that did not contain any metallic elements, and prove Heisenberg wrong? Theorists since Heisenberg, most notably McConnell, have devised mechanisms for ferromagnetic interaction between organic radicals containing only atoms from the first row of the Periodic Table. These cry out to be tested. Because organic radicals themselves are so chemically reactive, and the compounds difficult to stabilise and characterise, there have been several false alarms on the way to a true organic molecular ferromagnet.

The first such material to pass all the chemical and physical tests was synthesised only in 1991: it goes under the name of p-nitrophenyl nitronyl-nitroxide and had the less than spectacular T_c of 0.6 K. Still, as Dr Johnson said of the dog walking on its hind legs, the surprise is not that it is done well, but that it is done at all. Several other members of the nitronyl-nitroxide family also showed ferromagnetic interactions between neighbouring molecules, but so far without clear evidence of long-range order.

The compound with T_c of 1.48 K also contains nitroxyl groups, on which are located the unpaired electrons. But, whereas in the nitronyl-nitroxide series each molecule contains two N-O moeities connected by a single carbon atom as part of a five-membered ring, in the compound with the somewhat higher T_c, two N-O groups are deliberately built into the molecule in such a way that the orbitals carrying the unpaired electrons are orthogonal—just the recipe that was followed by the coordination chemists. Thus, even within the molecule, the electron spins on the two N-O groups should align parallel—and they do. More tricky to engineer is the packing of molecules in the crystal, so that the same effect occurs throughout the lattice. However, the susceptibility and magnetization results confirm that was achieved. Only one piece of behaviour typical of ferromagnets has not been observed, namely hysteresis. Hence (and perhaps not surprisingly), the coercive field is tiny, less than 0.1 Oersted.

What of the future? The dog having walked on its hind legs, albeit unsteadily, we can confidently expect T_c's of molecular organic magnets to increase further. Sticking my own neck out, though, I shall be surprised if a magnet containing only carbon, hydrogen, nitrogen and oxygen is ever made with T_c above room temperature. My reasons are twofold: first, the nature of organic radicals is that they contain only one unpaired electron in each repeat unit and, second, the interaction between the orbitals carrying the unpaired electrons takes place across distances equal to the van der Waals intermolecular distance. Both factors militate against high T_c's. Still, the search is fun and I shall be pleased to be proved wrong.

chapter 11

Mixed-Valence Compounds

When inorganic chemists had only their eyes to supplement the preparative dexterity of their hands, compounds containing the same element in two states of oxidation already occupied an important place in schemes of chemical classification. The bright colours of many of these materials had early attracted the attention of chemists both as curious physical phenomena and, quite frequently, also in view of the possible use of the compounds as pigments (Fig. 40 shows some typical examples). This early activity in the field of mixed-valence materials accounts for the variety of trivial names bestowed upon many of them: Prussian blue, tungsten bronze, Wolffram's red salt, and so on.

The significance of colour

Alfred Werner was probably the first to notice that deep colours frequently occurred when two oxidation states of an element were present. In 1896 he prepared a series of oxalato-platinites which, when stoichiometric, were lemon-yellow, but which readily oxidised to a copper sheen. In the 1960s these compounds were examined again in some detail by K. Krogmann, who found that whereas in $K_2Pt(C_2O_4)_2{\cdot}2H_2O$, the planar oxalato-complexes are well isolated from one another, in $K_{1.6}Pt(C_2O_4)_2{\cdot}2.5H_2O$ the complex anions are stacked one above the other so that the Pt-Pt distance is only 2.75 A. Werner drew an illuminating analogy between the intense colour of the

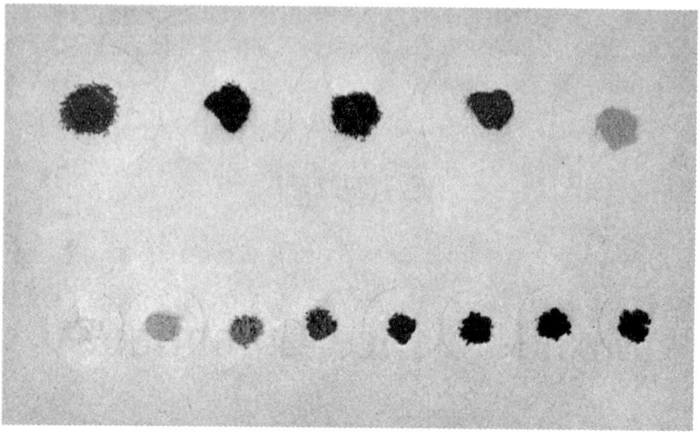

Fig. 40. Top row: a series of chlorocuprate(I, II) salts containing varying proportions of copper(I) and copper(II). On the far left and right respectively are the pure copper(I) and copper (II) salts. Note that the darkest sample contains roughly 50 per cent of each valence state. Bottom row: caesium hexachlorostannate(IV) samples doped with varying concentrations of hexachloroantimonate(III, V), $Cs_2Sb_xSn_{1-x}Cl_6$. The salt on the extreme right is Cs_2SbCl_6 and on the left Cs_2SnCl_6.

partly oxidised platinum compound and that of the purely organic compound quinhydrone, noting that both the fully reduced and fully oxidised derivatives of the latter (phenol and quinone) were colourless. He also made a comparison with the tungsten bronzes, and even noted that the latter contained a small proportion of a lower oxidation state in the presence of a higher, while in his own platinum compounds the reverse was true.

In the 25 years following Werner's observation, the idea slowly developed that light absorption could lead to oxidation and reduction, or electron transfer, between the same or different elements. Thus H. Wells wrote in 1922:

> *If it is assumed that the atoms of a metal in two states of valence in the same molecule, instead of retaining fixed individual valences, continually make these exchanges of electrons, it may be supposed that light, passing through such molecules, is in some way affected, so that colours or opacity are produced.*

In support of this hypothesis he cited the examples of gold(I, III), antimony(III, V) and copper(I, II) as chloride salts, and iron(II, III) as bromide and cyanide. The phrase 'constitutively coloured' was likewise used by K. A. Hofmann and K. Hoschele to describe solids whose colours were not simply the sum of the colours of their components, a situation most likely to occur when 'the oxidation states of two atoms can exchange under the influence of light'. They also touched on the important point that the deepest colours were found, not just when two atoms of an element were present in different oxidation states, but when the environments of the two atoms were similar. We shall return to this in more detail later.

A particularly neat experiment, demonstrating that the intensity of mixed-valence colour varied with the proportions of the two valence states, was performed in 1918 by the mineralogist T. L. Watson. When stoichiometric the mineral vivianite is colourless, and has the composition $Fe^{II}_3(PO_4)_2 \cdot 8H_2O$. Most samples, however, are pale blue owing to oxidation of some of the iron to Fe^{III}. That the colour was due to the Fe^{II} and Fe^{III} acting together, and not just to the Fe^{III} alone, followed from the behaviour of a microcrystalline sample on grinding in air. Watson found that the colour at first became dark (and analysis proved that the relative proportion of Fe^{III} had increased) but on further grinding it became greenish-blue, then greenish-yellow, and finally lemon-yellow when all the iron had been oxidised. Grinding the sample under an atmosphere of carbon dioxide of course produced no change.

It was initially suggested that in all mixed-valence compounds the 'valences' (that is, electrons) were in some kind of perpetual oscillation, called by Wells 'spontaneous electronic activity'. At about this time, however, E. Zintl and A. Rauch carried out an experiment which they held to be a refutation of the 'oscillating valence' concept. They prepared the orange-yellow mixed-valence compound $Pb_2O_3 \cdot 3H_2O$ using radioactive Pb^{II} and inactive Pb^{IV} in a way which they had previously shown did not lead to radioactive exchange. They then irradiated the compound with ultraviolet light for some hours, decomposed it, and found that only 1–2 per cent of the radioactivity had exchanged; they therefore concluded that no interchange of valences had occurred. Much more recently a similar experiment was performed on the $Sb^{III,V}$ compound Cs_2SbCl_6 (Fig. 40) with similar results. If, therefore, the light absorption were connected with some process of oxidation and reduction, this could only be taking place in the excited state and, moreover, the electron transfer excited state must have a high probability of reverting to the original un-oxidised or reduced ground state. As we shall see, such a situation in fact follows from an asymmetrical distribution of anions about the cations of differing valence.

Conductivity of mixed-valence compounds

Though unexpected colour was the first of their attributes to draw mixed-valence compounds to the attention of chemists, during the 1930s other less spectacular properties began to attract notice. The first was electrical conductivity. Following the establishment of the band theory of electronic states in solids, it became clear that there was a large class of materials, the transition-metal oxides which, according to the most straightforward interpretation of band theory, ought to have been metallic conductors, because the band formed by the interaction of the d-orbitals was only partially occupied, but which certainly were not. To resolve this dilemma, N. F. Mott pointed out that some critical inter-atomic separation should exist, outside which a molecular-orbital or band model is no longer suitable for describing the interactions between the cations. For large separations, a model which assigned a fixed integral number of electrons to each cation (the Heitler–London theory) would be more appropriate. This arises from the fact that if we have an ionic lattice

$$\cdots Ni^{2+}O^{2-}Ni^{2+}O^{2-}Ni^{2+}O^{2-}\cdots$$

a great deal of energy would be needed to create a polar state

$$\cdots Ni^{2+}O^{2-}Ni^{+}O^{-}Ni^{2+}O^{2-}\cdots$$

whereas the simplest form of molecular orbital wave-function gives equal weight to neutral and polar states. Because the cations interact only weakly, materials such as MnO and NiO are therefore semiconductors with a high activation energy and a very high resistivity at room temperature (approximately 10^{14} ohm cm).

However, it was quickly noticed that when they deviate from strict stoichiometry, transition-metal oxides may become very much better conductors than this. For example, green NiO can be oxidised to a black material containing about 0.5% excess oxygen, when the resistivity falls by ten orders of magnitude. The same effect can be obtained by heating stoichiometric NiO with lithium carbonate at 800°C, when a variable number of Ni^{2+} ions in the rock-salt lattice are replaced by Li^{+}, and a corresponding number of Ni^{2+} ions are oxidised to Ni^{3+}. There can be few physical properties of any material which can be varied so easily by a factor as large as that.

Mixtures of valence states can be induced in a great many other transition-metal oxide lattices by doping them with ions of fixed but different valence, and many highly conducting solids have thus been produced. Some of the combinations are shown in Table 1. They were called 'controlled-valence semiconductors' by E. J. W. Verwey and his colleagues at the Philips Research Laboratories after World War II, and have been widely used as temperature-dependent resistors ('thermistors'). That the conductivity is definitely associated with the simultaneous presence of two valence states may be demonstrated in a few instances by preparing a complete set of isomorphous materials covering the whole range of proportions of the two valences. Figure 41 shows the resistivities of two examples, one an oxide, the other a halide. The mixed-valence materials all have much lower resistivities than either of the single-valence ones.

Mixed-valence alone, however, is a necessary, but not sufficient criterion for high conductivity in the transition-metal oxides, as the compounds M_3O_4(M = Mn, Fe, Co) show. All three are spinels, but the resistivity of Fe_3O_4 is at least ten orders of magnitude lower than that of the other two. The spinel structure is best thought of as a cubic close-packed lattice of oxide ions containing tetrahedral and octahedral holes, among which the divalent and trivalent cations are distributed. In 'normal' spinels, divalent ions occupy exclusively the tetrahedral, and trivalent ions the octahedral sites, $(A^{II})^{tet}(B^{III})_2^{oct}O_4$, but there also exist 'inverted' spinels in which

Table 1. Some examples of controlled-valence semiconductors.

Host lattice	Crystal structure	Doping substance	Valence induced
NiO	rock-salt	Li_2O	Ni^{III}
TiO_2	rutile	Ta_2O_5	Ti^{III}
$ZnFe_2O_4$	spinel	TiO_2	Fe^{II}
$LaMnO_3$	perovskite	SrO	Mn^{IV}
$CaTiO_3$	perovskite	La_2O_3	Ti^{III}

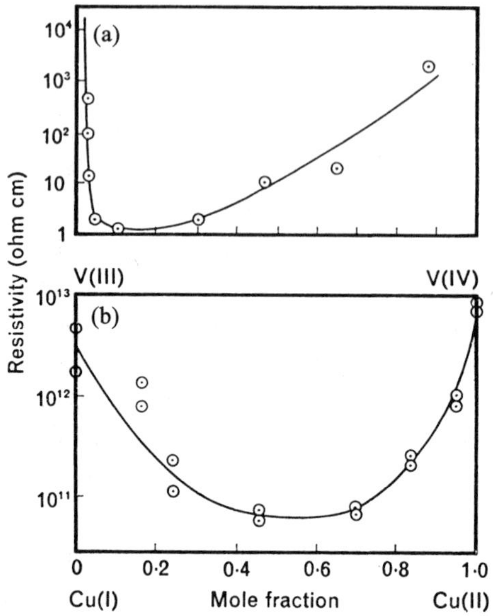

Fig. 41. The resistivities of some mixed-valence compounds of varying composition: (a) solid solutions of $MgV_2^{III}O_4$ and $Mg_2V^{IV}O_4$; (b) hexamminecobalt(III) chlorocuprates (I,II).

all the divalent and half the trivalent ions are found in the octahedral sites and the tetrahedral sites are occupied by the remaining trivalent ions, $(B^{III})^{tet}(A^{II}B^{III})^{oct}O_4$. Both Mn_3O_4 and Co_3O_4 are normal spinels so the atoms of differing valence occupy distinguishable sites, in contrast to Fe_3O_4 in which both Fe^{II} and Fe^{III} are tetrahedrally coordinated. Hence, just as Hofmann and Hoschele pointed out for the colours of mixed-valence compounds, the highest conductivities occur when the two kinds of ion occupy sites with very similar geometries.

The equivalence of two sites in a lattice is determined not only by the geometrical distribution of ions around them, but also by the orientation of any unpaired electron spins on the ions occupying them. Mixed-valence leads to many striking magnetic effects in solids which can often be correlated with conductivity. Possibly the best examples of such a correlation are the series of manganites also first investigated at the Philips Laboratories. $LaMn^{III}O_3$ and $SrMn^{IV}O_3$ are both perovskites, the former somewhat distorted by the Jahn–Teller effect, the latter cubic. The Mn^{III}-O-Mn^{III} magnetic interaction is weakly ferromagnetic along two axes and antiferromagnetic along the third, while Mn^{IV}-O-Mn^{IV} wholly antiferromagnetic; both compounds, as might be expected, are high-resistance semiconductors with high band-gaps, like stoichiometric NiO. When La and Sr are both present in the lattice, on the other hand, the resulting mixed valence of the Mn ions leads to strong ferromagnetism and metallic conductivity. Indeed, plots of the variation of the Curie temperature and the conductivity with composition are strikingly similar, as Fig. 42 shows. The asymmetrical position of the maximum in the Curie temperature-composition diagram is a result of superimposing the $Mn^{III,III}$ and $Mn^{IV,IV}$ interactions on the strongly ferromagnetic $Mn^{III,IV}$ interaction.

Another example of a ferromagnetic mixed-valence compound is, of course, magnetite, Fe_3O_4. Here, neutron diffraction experiments show that while the Fe^{II} ions in the tetrahedral sites are weakly coupled in an antiferromagnetic sense to both the Fe^{II} and the Fe^{III} in the octahedral sites, the latter have a much stronger ferromagnetic interaction with each other. Measurements of the Mössbauer effect also confirm that electron transfer takes place between the octahedral sites and it is here therefore that conductivity arises. At room temperature the magnetic fields at the octahedral sites are indistinguishable, indicating that electron transfer between the Fe^{II} and Fe^{III} is taking place much faster than the time-scale of the Mössbauer experiment, that is, faster than 10^8 per second. At 85 K on the other hand, Fe^{II} and

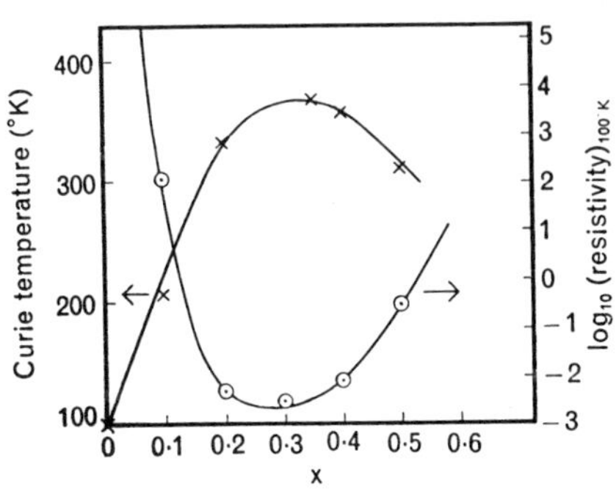

Fig. 42. The Curie temperatures and resistivities of lanthanum strontium manganites(III, IV).

Fe^{III} ions are separately distinguished, an observation which complements Verwey's original work on the temperature dependence of the conductivity. He found that below 120 K the conductivity of magnetite fell suddenly by two orders of magnitude, and the crystal lattice became anisotropic.

Accurate neutron diffraction work has since revealed a phase transition at this temperature: above 120 K the octahedral Fe-O distances are uniformly 2.059 A whereas in the low-temperature phase there are two octahedral sites, one (presumably the Fe^{III}) having an Fe-O distance of 2.00 A and the other ($Fe^{III}O$) 2.12 A. Thus above 120 K the valences are indeed in rapid oscillation, while below, they are more or less firmly trapped.

A connection between magnetism and electrical conductivity is implied in the super-exchange mechanism proposed by P. W. Anderson to account for the interaction between unpaired electrons on neighbouring cations which are separated by anions, like those in magnetite and the manganites. The leading term of the exchange integral contains a product of atomic wave-function of the anion and the two cations which suggests that, in the manganites for example, one should visualise the process of electron transfer

$$Mn_A^{III}O^{2-}Mn_B^{IV} \rightarrow M_A^{IV}O^{2-}Mn_B^{III}$$

as the transfer of an electron from O^{2-} to Mn^{IV}, accompanied simultaneously by transfer of another electron from Mn^{III} to the resulting O^{-}. Hund's rule predicts that the O^{2-}electron whose spin is parallel to those in the d-shell of the Mn^{IV} will be transferred most easily, leaving behind an electron of opposite spin. Then the electron from Mn^{III} which is transferred into the resulting hole on O^{-} must have the same spin as the one transferred from O^{2-} to Mn^{IV}. Hence the net result of the electron transfer is a ferromagnetic coupling between the Mn^{III} and Mn^{IV} electrons, just as observed.

To obtain the strongest ferromagnetic interaction, as well as the highest conductivity, by this kind of mechanism, a valence interchange between the cation sites must be accomplished with minimum expenditure of energy, a situation which will occur when, as in the manganites, the sites are geometrically identical. If, as in Fe_3O_4 below 120 K, there are two cation sites (A and B, say) with distinctly different metal-oxygen distances, the energies of configurations

$$Fe_A^{III}O^{2-}Fe_B^{II} \text{ and } Fe_A^{II}O^{2-}Fe_B^{III}$$

are clearly different, and even if the ionisation potential of Fe^{II} were equal to the electron affinity of Fe^{III}, vibrational energy would still be required to inter-convert them. In the absence of that energy the valences would be effectively 'trapped'.

So far we have concentrated on what might be called the 'unusual' properties of mixed-valence compounds, that is, those which are not simply the sum of the atomic or molecular properties of the constituents, for example, dark colour, ferromagnetism or metallic conductivity. However, there are many mixed-valence compounds which show none of these peculiarities. A good example is $GaCl_2$, a colourless diamagnetic insulator. The structure of this compound provides a clue to its lack of collective properties: two types of gallium site are found, one surrounded tetrahedrally by chlorine atoms at 2.19 A, the other an irregular dodecahedron with Ga-Cl distances of 3.2 A. The compound should clearly be formulated $Ga^{I}(Ga^{III}Cl_4)$, and in support of such a formulation the Raman and far-infrared spectra contain bands at frequencies very close to those of the $(GaCl_4)^-$ ion in solution. And though it has not apparently been determined, we would also expect the ultraviolet spectrum of $GaCl_2$ to be just that of $(GaCl_4)^-$ and Ga^+, in the same way that in a mixed-valence compound like $(Co^{III}(NH_3)_6)_2(Co^{II}Cl_4)_3$ we would find a simple superposition of the electronic spectra of $(Co(NH_3)_6)^{3+}$ and $(CoCl_4)^{2-}$.

Classifying mixed-valence compounds

It was this obvious connection between electron delocalisation in mixed-valence compounds and the geometrical similarity between the sites occupied by the different cations that led M. B. Robin and I to propose a general classification scheme embracing all such materials. Our scheme was derived by considering the factors that determine the wave-functions of the valence-shell electrons. For simplicity, let us consider a pair of metal ion sites A and B such as those we have described in the low-temperature form of magnetite, and suppose that one site is occupied by Fe^{II}, the other by Fe^{III}. The zero-order wave-function of the ground state is simply the product $\phi_A^{II}\phi_B^{III}$ but there are also excited states $\phi_A^{III}\phi_B^{II}$ in which an electron has been exchanged between the two sites. Some of these excited states may have the same symmetry as the zero-order ground-state wave-function, and hence may mix into it to an extent determined by the variation principle, to give a true ground-state wave-function

$$(1-\alpha^2)^{1/2}[\phi_A^{II}\phi_B^{III}]+\alpha[\phi_A^{III}\phi_B^{II}].$$

The mixing coefficient α depends, among other things, on the energy of the configuration $Fe_A^{III}Fe_B^{II}$ compared with $Fe_A^{II}Fe_B^{III}$, which again depends on how different the environments of A and B are.

Three categories of behaviour may therefore be distinguished. If the environments of sites A and B are very different (for example, one is surrounded tetrahedrally by chloride ions, the other octahedrally by water molecules) then the ligand field stabilization energy of an Fe^{II} ion placed at one site or the other will also differ greatly. The energy of $\phi_A^{II}\phi_B^{III}$ then far exceeds that of $\phi_A^{III}\phi_B^{II}$ and the ground state is described to a very good approximation as $Fe_A^{II}Fe_B^{III}$. There will be no electrical conductivity, and the light absorption will be the sum of Fe^{III} in A and Fe^{II} in B. This rather uninteresting situation we call Class I: $GaCl_2$ is a good example.

At the opposite extreme, if the sites A and B are indistinguishable, the energies of the two configurations $Fe_A^{II}Fe_B^{III}$ and $Fe_A^{III}Fe_B^{II}$ are equal and the wave-function of the ground state contains equal contributions from each of them, that is $\alpha = 2^{-1/2}$, a situation we call Class III. In this case, we could visualise the 'extra' electron moving in a uniform potential created by the oxidised cations, that is, instead of writing $Fe_A^{II}Fe_B^{III}$ one might write $Fe_A^{III}Fe_B^{III} + e^-$. If they formed infinite continuous lattices, such compounds would be metallic conductors and would have none of the single-ion properties, such as electronic, vibrational or Mössbauer spectra characterising the metal in either of its oxidation states. The mixed-valence manganites discussed above, and magnetite at temperatures higher than 120 K, are both excellent examples of Class III mixed-valence compounds, though probably the best known of all are the tungsten bronzes.

In some mixed-valence compounds the metal ions, although sited in identical environments, do not form continuous lattices but discrete clusters. Examples of this kind are the niobium and tantalum chlorides Nb_6Cl_{14} and Ta_6Cl_{14}, which contain cubic $(M_6Cl_{12})^{2+}$ ions with an average metal oxidation state of 2.33. As there is no direct bridging between the clusters in the crystal, such materials do not behave as metals or even semiconductors. Neither, on the other hand, do they exhibit any single-ion properties. Thus we divide Class III materials into two groups: infinite lattices and cluster compounds.

Between our two extreme Classes I and III lie a vast number of other interesting materials in which the sites occupied by the ions differing in valence are similar though still distinguishable. For example the sites might both be octahedral, but have different metal-ligand distances, as in magnetite below 120 K, or have different ligand atoms, as in Prussian blue. We then expect values of α appreciably greater than zero, though less than $2^{-1/2}$. Because the electron delocalisation is then small, 'single-ion' properties still persist in the mixed-valence system, often with very little perturbation. For example, in the chlorocuprates(I, II) in Fig. 40 the ligand field absorption bands of Cu^{II} appear in the spectrum of the mixed-valence salt, while in the chloroantimonates(III, V) illustrated in the other half of this figure the vibrational frequencies of the constituent ions $(Sb^{III}Cl_6)^{3-}$and $(Sb^{V}Cl_6)^{-}$ are practically unchanged. Nevertheless, if two configurations such as $Cu^{I}_{A}Cu^{II}_{B}$ and $Cu^{II}_{A}Cu^{I}_{B}$ are no longer quite orthogonal in this class of compounds, optical transitions can take place between them, and it is to these that we owe colours like those in Fig. 40. The absolute intensities of the mixed-valence bands lead directly to estimates of α; values derived for Prussian Blue and hexachloroantimonates(III, V) are around 0.01. The band frequencies measure the energy required for electron transfer between the two sites, but electrons may also be transferred by thermal excitation, so that the Class II compounds like those in Fig. 41 are semiconductors.

Table 2 assembles some predictions about the physical properties of all the classes we have defined. From a knowledge of the structure, and hence class, it should be possible therefore to predict roughly what physical property any mixed-valence material might have or, conversely, from the known properties to make an intelligent guess at the structure. Clearly, such predictions cannot be made with complete accuracy; that is a task for the future.

This brief survey will have shown that mixed-valence compounds possess an astonishing variety of physical and chemical behaviour, and that

Table 2. Physical properties of the classes of mixed-valence compounds.

	Class I	Class II	Class IIIA (clusters)	Class IIIB (infinite lattice)
Optical	No inter-valence transitions in visible	One inter-valence transition in visible; low intensity	One or more intervalence transitions in visible; high intensity	Opaque; metallic reflection in visible
Electrical	Insulator; resistivity greater than 10^{12} ohm cm	Semiconductor; resistivity 10–10^{8} ohm cm	Insulator or high-resistance semiconductor	Metallic conductor; resistivity 10^{-2}–10^{-9} ohm cm
Magnetic	Diamagnetic or paramagnetic to very low temperature	Magnetically dilute; either ferro- or anti-ferromagnetic at low temperatures	Magnetically dilute	Pauli paramagnetism or ferromagnetic high with high Curie temperature
Other properties (IR, Moessbauer etc.)	Spectra of constituent constituent ions at normal frequencies	Spectra of constituent ions close to normal frequencies	No spectra of constituent ions	No spectra of constituent ions

they pose fundamental problems in the inter-relation of spectroscopy, electronic conductivity and magnetism. I have tried to show how a simple model based on ionic site symmetries can serve at least as a classification scheme, and help in interpreting some of the spectacular properties of these substances, which have fascinated chemists for most of the twentieth century.

chapter 12

Superconductors Past, Present, and Future

Not often does a discovery in basic science reach the front pages of the daily newspapers, but towards the end of March 1987, papers throughout the world carried a series of remarkable headlines: 'Electric Dream Has Come True', 'Scientists Herald Electronic Revolution', and 'The Woodstock of Physics' were just a few of them. Even more extraordinary was the fact that the event occasioning the headlines was a scientific conference, the kind of gathering of professionals not usually reported outside the staid pages of academic journals. What brought the world's TV and newspaper reporters to the Spring Meeting of the American Physical Society in Boston was an announcement that a series of novel chemical compounds had been discovered, which lost all electrical resistance at a much higher temperature than any previously known. The property of conducting electricity without resistance is called superconductivity, and the materials in question were dubbed high temperature (or, for short, high T_c) superconductors.

My purpose here is to fill in the background that led up to what can only be called the hysteria that gripped the scientific world in 1987. I shall do so, first by exploring what superconductivity is and how it was first discovered, and then by describing how the high T_c materials came to be found, and the detective work needed to uncover their chemical composition and structure. Being at heart (as well as by training) a chemist, I shall also give a glimpse of the rich variety of other substances exhibiting this remarkable property, some of which (who knows?) may lead to advances as dramatic as those of 1987. The drama, of course, would not only be for science. The notion that electricity might be transported over long distances without expending energy is one to grip the imagination of engineers and technologists: that is why the news from the American Physical Society meeting made front page headlines. It is therefore pertinent to ask how far those dreams have been fulfilled. First, though, we must go back nearly two hundred years, to consider how electricity is conducted by metals, and what the prefix 'super' means in practice. Only then can the discoveries of the 1980's be placed in their proper context.

Perfect conductors and superconductors

One of the most important accomplishments of my distinguished predecessor as Director of the Royal Institution, Sir Humphry Davy, was to isolate several elements in groups 1 and 2 of the Periodic Table as a result of his work in electrochemistry. In fact, I like to tell visitors that about seven per cent of all the non-radioactive elements first saw the light of day at 21 Albemarle Street. Davy summarised many of his results in his Bakerian Lecture to the Royal Society in 1807 on 'The decomposition and composition of the fixed alkalis'. After describing in that lecture some of the physical properties of what he called potash (but we would call potassium) Davy adds the bold statement: 'it is a perfect conductor of electricity'. Was he right? Sadly, no. Potassium is, of course, a metal and one of the better conducting metals at that, with a specific resistivity at room temperature of 60 micro-ohm cm. But that is certainly not zero.

Indeed, much later in his career Davy returned to the general question of the conductivity of metals in a paper on 'The magnetic phenomena produced by electricity', published in the Philosophical Transactions of the Royal Society in 1821. Having determined, as he put it, that 'there was a limit to the quantity of electricity which wires were capable of transmitting', he conducted an exhaustive set of experiments on a variety of metals, designed to find out how the 'conducting power' depended on temperature, surface area, length, and (remarkably in view of its relevance to superconductivity) magnetic field. Summarising the results of his experiments,

Davy wrote: 'The most remarkable general result that I obtained by these researches, and which I shall mention first, as it influences all the others, was that the conducting power of metallic bodies varied with the temperature, and was lower in some inverse ratio as the temperature was higher'. What he had observed, therefore, was the general characteristic of metals that their conductivity decreases with increasing temperature.

Given that result, and with the benefit of hindsight, we can re-approach Davy's 1807 work on potassium and enquire if he might have been right if the temperature was lowered sufficiently. Again, sadly, the answer must be no. Fig. 43 shows the resistance of a sample of potassium measured at very low temperature: the resistance indeed decreases as the temperature is reduced but below 5° above absolute zero (5 K) it becomes more or less constant at about five per cent of its value at room temperature. We know now that the electrons carrying the current, which behave as waves, are scattered when the atoms in the crystal lattice do not form an ideally equally

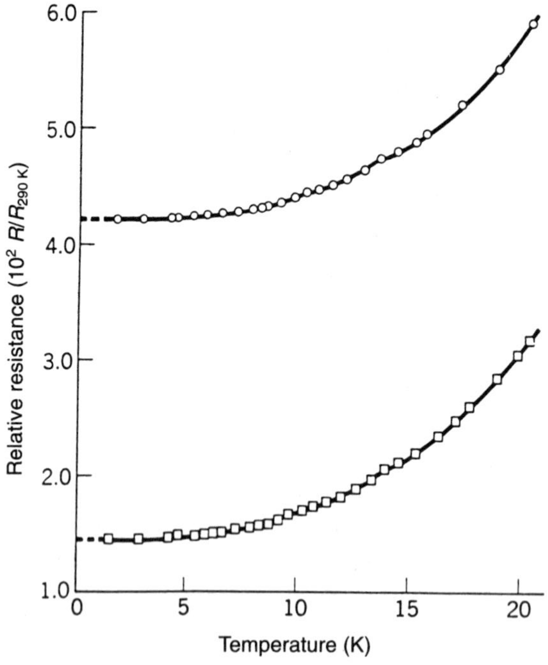

Fig. 43. The resistance of two samples of potassium at low temperature.

spaced array. Such a situation can arise for two reasons. One is that real solids always contain impurities or occasional vacant lattice sites: the lower of the two curves in Fig. 1 was obtained from a purer sample than the upper one. Secondly, at any temperature higher than absolute zero the crystal lattice vibrates, so if we take an instantaneous photograph of the crystal, the atoms will not be found on their ideal lattice sites. The latter is the physical origin of the phenomenon observed by Davy: as the temperature rises, the lattice vibrates more, the electrons are scattered more, and the conductivity goes down.

After Davy's work the world had to wait another century before the first truly perfect conductor was found and, as is so often the case in science, it was an unexpected consequence of an apparently unrelated development. Throughout the 1890s and into the beginning of the present century a polite but intense professional rivalry existed between the laboratories of the Royal Institution and the Physics Department of the University of Leiden in The Netherlands, personified by their two directors, respectively Sir James Dewar and Kammerlingh Onnes. Both were engaged in liquefying the rare gases, which required the production of very low temperatures. Dewar succeeded in every case except for helium, a final prize that fell to Onnes in 1908. It was therefore he, and not Dewar, who was in a position to start examining the properties of materials cooled to the lowest temperatures available at that time anywhere in the world. In view of the important part played by impurities and temperature in the conductivity of metals, Onnes decided to measure the resistance of a highly pure metal cooled to very low temperature and, because it could be purified by distillation, he chose mercury: the result (shown in Fig. 44, taken from his original publication on the subject) was extremely surprising. The resistance decreased in the way characteristic of a metal as the temperature was lowered towards 4.2 K, but suddenly at about 4.3 K it dropped abruptly, becoming (as far as he could measure it) zero. The first perfect (or, as we now call it, super) conductor had been found.

Onnes's picture (Fig. 44) gives us a clue to several important aspects of superconductivity. First, the electrical resistance is not just low, it really is zero: current has remained circulating in suitably cooled superconducting wires for years without detectable degradation. Second, for every metal that exhibits it, the phenomenon occurs only below some characteristic

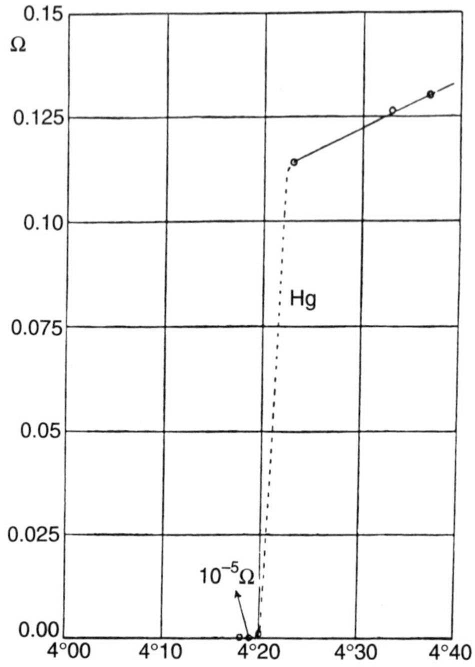

Fig. 44. The resistance in ohms of mercury at low temperature, as measured by Kammerlingh Onnes.

temperature called T_c; for some metals (such as potassium, as it happens) it does not occur at all, even at the lowest temperatures accessible nowadays, numbered in millikelvins. A third matter, not apparent from Fig. 44, is that a superconductor is a perfect diamagnet. It was actually Faraday who first divided all substances into two classes, paramagnetic and diamagnetic, according to whether lines of magnetic flux (he called them 'magnetic conduction') arising from a uniform magnetic field bathing a lump of the material were drawn into it or expelled from it. The two cases are shown in Fig. 45, in a diagram taken from Faraday's paper, published in his Experimental Researches in Electricity in 1852. In fact, a superconductor is a perfect diamagnet, with no lines of flux inside at all, in contrast to ordinary metals, which are weakly paramagnetic. However, that is only the case for magnetic fields below some critical value, H_c, above which the material no longer superconducts, but reverts to a normal metal. Clearly, for any

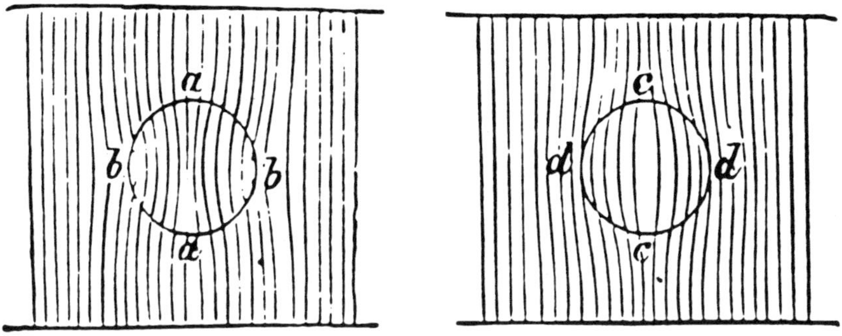

Fig. 1 Fig. 2

Fig. 45. Lines of magnetic conduction (*sic*) around a paramagnetic and a diamagnetic specimen, as drawn by M. Faraday, labelled respectively 'Fig. 1' and 'Fig. 2'.

practical use that one may wish to make of superconductivity, H_c is equally important as T_c.

Below T_c and H_c, our material is in quite a different state from that above these values; really a new state of matter. This is not the place to expound the theoretical explanation for the zero resistance and perfect diamagnetism, which was only established some 40 years after the original discovery, and is associated with the names of Bardeen, Cooper, and Schrieffer (called BCS theory for short). Suffice it to say that in a normal metal the electrons move through the crystal lattice essentially independently of one another. By contrast in a superconductor they move as pairs (so called 'Cooper pairs'). Agitating the lattice by heating it up, or applying a magnetic field, are two ways of pulling these pairs apart so that they revert to independent motion.

Exploiting superconductivity

The notion that wires can be made that would carry an electric current without any resistance is one to make an electrical or electronic engineer dream. But the dream can only be realized by keeping the wire at a temperature below T_c. Until a few years ago, the only way to achieve the temperatures needed was to use liquid helium, still (95 years after its first liquefaction by Onnes) an exotic and expensive fluid. Not only is helium quite a rare element, produced in nature as a by-product of radioactive decay of heavy elements and hence found only in a few isolated deposits, but it has to be stored and transported in elaborately insulated containers. The latter, ironically, are called 'dewars'. Still, despite the complications and expense, superconductivity is such a unique property that it has found quite a number of applications, albeit of a specialized 'high tech' kind.

Many of the present day applications hinge on making magnets. In the Royal Institution it is scarcely necessary to dwell on the fact that when a current is passed through a wire formed into a coil, a magnetic field is produced at the centre; the bigger the current, the bigger the field, but in a normal metal the wire gets hot. Along a superconducting wire one can pass extremely large currents without it heating up provided only that the magnetic field generated by the coil remains below H_C. A coil made by Faraday, with about 200 turns on a cardboard tube, should produce a field of about 2000 G; one of comparable size made from a superconducting niobium alloy can reach about 30 times higher.

Very high field magnets first found application in physics research laboratories but, with the invention of nuclear magnetic resonance, which found wide acceptance as an analytical tool, they quickly came to be found in chemistry and biology laboratories. Companies such as Oxford Instruments sprang up to manufacture these complex and delicate items of equipment (Fig. 46). Even more widespread nowadays are the magnetic resonance body scanners found in many hospitals, based on superconducting magnets large enough to insert a human frame into. What is believed to be the largest superconducting magnet ever constructed is shown in Fig. 47. It was built at the Rutherford Appleton Laboratory (RAL) for installation at the Large Electron-Positron collider (LEP) at the CERN High Energy Particle Physics Laboratory in Geneva. But even so large a solenoid has to be filled up inside with liquid helium. Another important field of application for superconductors is to make extremely sensitive detectors of minute variations in magnetic fields. When found in the earth's crust such changes might signal hidden oil or mineral deposits or, in the human body, what we might call 'brain waves'. The detectors in question are called SQUIDS (superconducting quantum interference devices), and consist of two layers of superconductor separated by an extremely thin layer of an insulator. Roughly speaking, the ability of electrons to travel through the insulator from one superconductor to the other is affected by small changes in the

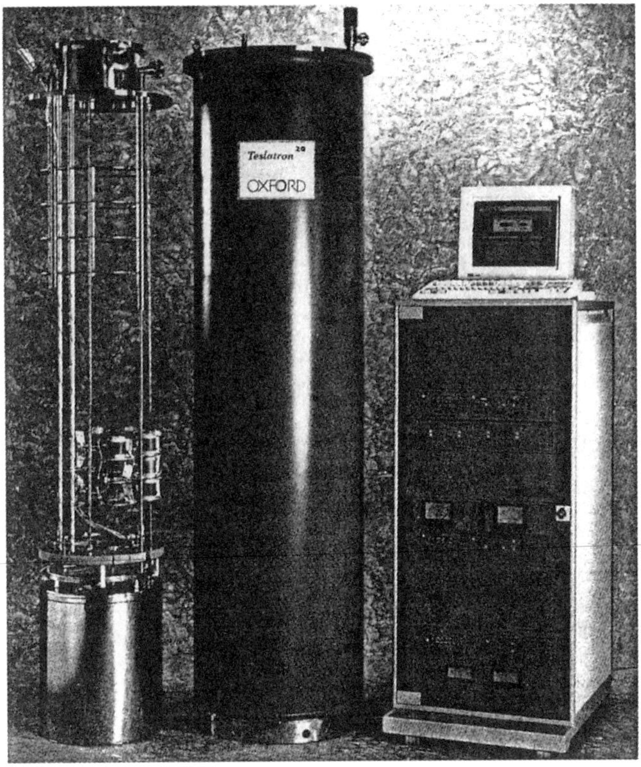

Fig. 46. A high field superconducting magnet and its cryostat. (Photograph courtesy of Oxford Instrument Co.)

magnetic field of the surroundings. It is against this background of important technology, hampered by the need to use elaborate cooling devices and expensive liquid helium, that the events of 1987, epitomized by the newspaper headlines I quoted, have to be seen.

1987: Annus mirabilis for superconductors

After Kammerlingh Onnes's unlooked-for discovery of superconductivity in 1910, more and more metals were found to have this property at higher and higher temperatures. The league table of T_c's was headed successively by other elements, culminating in niobium at 9.3 K in 1930. After that, binary

Fig. 47. The largest superconducting magnet in the world.

alloys came into the picture and occupied metallurgists and materials scientists for many years, so that by the early 1970s the highest known T_c had crept up to 23 K. And there matters rested for 15 years. Learned theoretical papers were written predicting that T_c could never go higher than 25–30 K, and it is said (though I have not checked) that at that time candidates taking the theory option in the Cambridge Physics Tripos examination were asked from time to time to calculate the maximum T_c attainable under the BCS regime.

What a surprise then when, at the end of 1986, a short article appeared in the German physics journal *Zeitschrift für Physik* under the unassuming title 'Possible high T_c superconductivity in the La-Ba-Cu-O system'. The article was written by two Swiss scientists from the IBM Laboratory at Ruschlikon near Zurich, George Bednorz and Alex Müller. They used the cautious word 'possible' because they had detected zero resistance but not, up till then, the other signature of superconductivity, diamagnetism. To a chemist, they also showed commendable caution in not identifying the formula of the compound concerned—in fact they did not know it. But the important measure was T_c; it was 30 K. After 15 years T_c had started to climb again.

Apart from the high T_c, the other important characteristic of this new system was that it was not an alloy of the traditional kind known to metallurgists, but a ceramic. Oxide superconductors had been known prior to 1986 but with relatively low T_c (e.g. $SrTiO_{3-x}$, 0.3 K; $BaPb_{0.7}Bi_{0.3}O_3$, 13 K; $LiTi_2O_4$, 13 K). Whatever the formula and structure of the new compound, the fact that it was an oxide placed it firmly within the realm of inorganic chemistry. Chemical analysis combined with X-ray diffraction, compared with measurements of the amount of diamagnetic material in specimens made by various methods in Europe, the USA, and Japan, showed very quickly that the optimum composition for superconductivity was $La_{1.85}Ba_{0.15}CuO_4$.

Consider this formula a bit more closely: if the compound had contained no Ba (always considered divalent by chemists), but only La (usually trivalent), and if oxygen had its normal valence of two, the copper would also have its customary di-valence. Such a compound, La_2CuO_4, exists and is not a superconductor: it is not even metallic. On replacing a small fraction of the La^{3+} by Ba^{2+}, however, it becomes metallic and (for a fraction of Ba between about 6 and 10 per cent) superconducting. Under these circumstances the copper no longer has an integral oxidation state, but an average of say 2.15. Its valence shell contains a number of electrons that is neither integral nor a rational fraction. From work that we and others had done in the 1960s and 1970s (see previous chapter), that kind of situation (called 'mixed-valence' by solid state chemists) does lead to metallic conductivity, though it was not obvious why such a compound should superconduct. At once we decided to look into the finer details of the crystal structure and lattice vibrations, believing that these would hold the key to the mechanism.

At that time, the ceramic methods of preparation did not permit single crystals to be grown, so we worked with microcrystalline samples which, nevertheless, were beautifully crystalline over small volumes, as may be seen from the electron micrograph in Fig. 48. Most important to us was

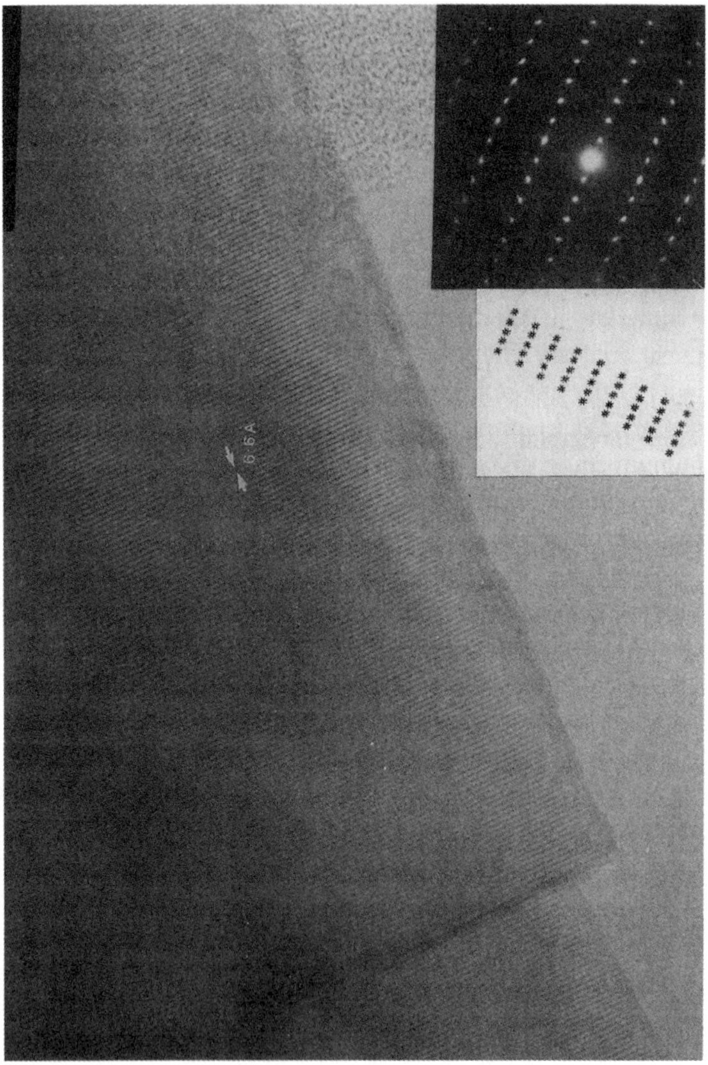

Fig. 48. Electron micrograph of $La_{1\cdot 85}Sr_{0.15}CuO_4$. The planes of Cu and O are separated by 6.6 A.

the precise positioning of the oxygen atoms around the copper, since that was the main factor determining the energy distribution of electrons in the conduction band. However, since they contain fewer electrons, oxygen atoms scatter X-rays much less than the other constituents of the crystal, hence diffraction experiments are best conducted using neutrons. On a wet Sunday afternoon in March 1987 we therefore found ourselves at the Rutherford Appleton Laboratory (RAL) at Chilton on the Berkshire Downs, in a small hut that houses the sample mounting and detectors of HRPD, the high resolution powder diffractometer. Some of the first data we obtained are shown in Fig. 49. A reporter from *The Times* got wind of the fact that we were working on this novel material and interviewed me, leading to my only appearance on the front page of a national newspaper. The reason for the sudden press interest in our rather technical experiments was that news was emerging of an even more dramatic increase in T_c that was to change the landscape of superconductivity irrevocably.

It had just been found, rather surprisingly, that applying pressure to $La_{1.85}Ba_{0.15}CuO_4$ increased its T_c quite noticeably. A group at the University

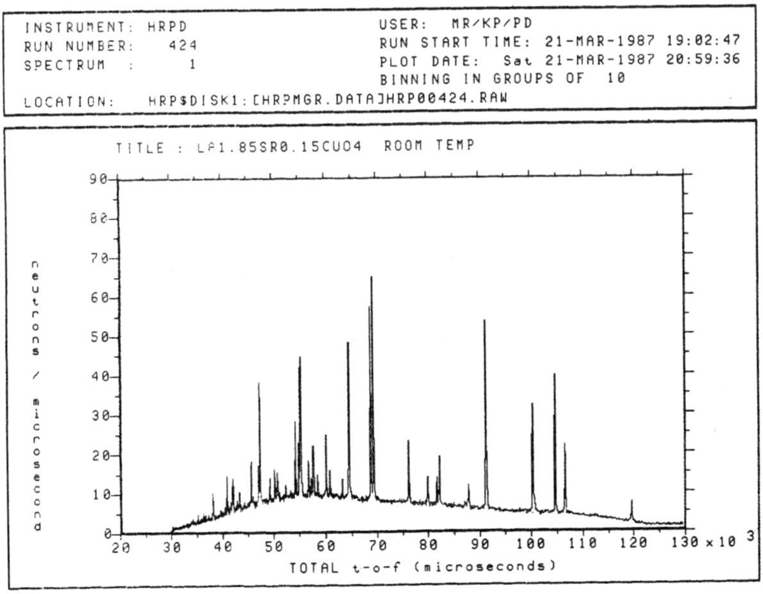

Fig. 49. Powder neutron powder diffraction profile of $La_{1.85}Sr_{0.15}CuO_4$ original data recorded on 21 March 1987.

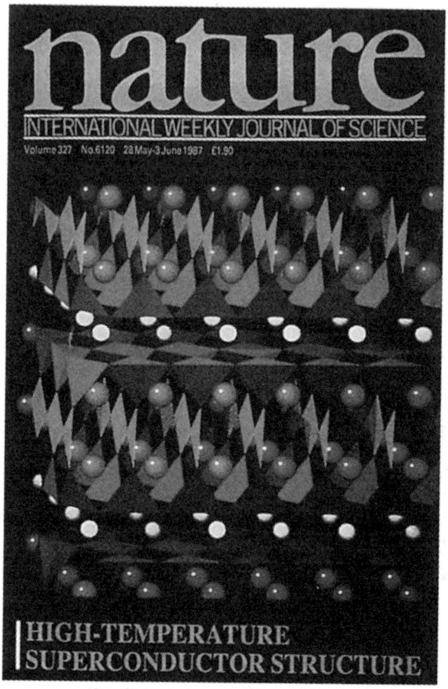

Fig. 50. The crystal structure of $YBa_2Cu_3O_7$.

of Houston led by Paul Chu therefore decided to make a new compound by replacing La with the smaller Y. And lo and behold, T_c went up to more than 80 K! Ironically, like Bednorz and Müller, Chu did not know the formula of the compound his group had made, but a world-wide chase led quickly to the formula being established. The compound in question turned out not to be the analogue of the La, Ba one but had quite a different composition: $YBa_2Cu_3O_{7-x}$. Its structure too is quite different. The RAL group obtained a small powder sample and collected neutron diffraction data and we published a joint paper in *Nature* (Fig. 50). This was the material that led to the newspaper headlines with which I started.

The era of high T_c superconductors

When its composition is fully optimized, the material whose structure is shown in Fig. 50, variously called YBCO, or 123 for short, has a T_c of 93 K.

The boiling point of liquid nitrogen is 77 K so, potentially at least, exploitation of superconductivity in electrical and electronic engineering could be transformed. Not only is liquid nitrogen much easier to store and transport than liquid helium, it is much cheaper. In fact it used to be said that whereas liquid helium costs about as much as Scotch whisky, liquid nitrogen costs about the same as milk. But there are difficulties.

As already noted, the high T_c superconductors are not alloys but ceramics, a fact that has profound consequences for the way in which they are synthesised and fabricated. With regard to synthesis, conventional ways of making ceramics by heating mixtures of oxides together for long periods at high temperatures are certainly not conducive to forming material that can be made into wire. Also, the main characteristic of ceramics is brittleness and the high T_c compounds are no exception: a lump of YBCO has about the same mechanical strength as a tea cup! Grain boundaries, dislocations, and other defects also have an important effect on the key superconducting parameters such as T_c and H_c and also the critical current density. Much ingenuity has been devoted to improving these properties, with results that may be seen from Fig. 51.

Whilst it is not possible to make wires out of ceramics alone, the friable and rather brittle material can be encapsulated in thin-walled silver tubes, which are then extruded at high temperature so as to trap the grains. Alternatively, it has been rolled into layers between sheets of silver and then cut up to form tapes, which can then be rolled into magnet coils. For electronic purposes, thin films have been deposited from the vapour phase by evaporating solid material, bombarding it with an intense laser beam. Another technique favoured by ceramicists for forming thick films is sol-gel processing. Organic salts containing the necessary Y, Ba, and Cu are hydrolysed to form a glassy homogeneous layer, which is then fired in a furnace to decompose it into the desired oxide.

Not only has a huge effort been made to optimise the properties and fabricate usable samples of YBCO but, in parallel, solid state chemists have continued their search for related materials that may have even more desirable properties. For example, about one year after the discovery of YBCO, Japanese workers prepared a series of phases in which the Y was replaced with Bi and the Ba with a mixture of Ca and Sr. Like the original La, Ba compound and YBCO, the Bi compounds (given the acronym BISCCO) contain

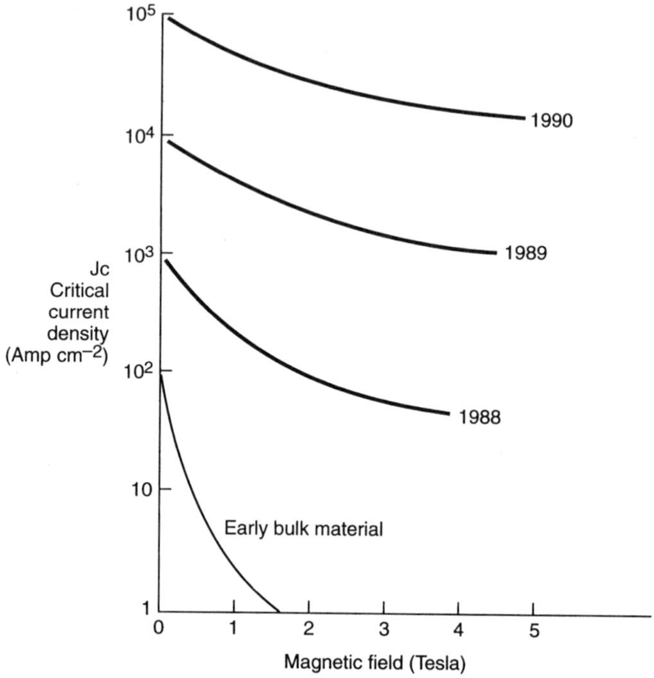

Fig. 51. Improvement in critical current density of $YBa_2Cu_3O_7$ samples from 1987 to 1990.

layers of corner-sharing CuO_4 units, but present two extra features that are novel. First, there is not one but a series of compounds with different numbers of CuO layers sandwiched between layers of Bi and O, which gives each one a different T_c. Second, the BiO and CuO layers are not precisely flat, but have wavy corrugations. The electron microscope image in Fig. 52 shows a view parallel to the layers, clearly revealing the modulation which is brought about by a mismatch between the preferred Bi-O and Cu-O bond lengths. Since, of necessity, they have to fit together in the lattice, a compromise is established in which first one and then the other becomes longer and shorter. This is an example of what physicists call 'competing interactions'. If the energy of the system is optimum for a variation in the lattice spacing that is not a rational fraction of the lattice repeat distance, the wavelength of the wavy corrugation (or modulation) is not an integral number of lattice spacings. Such structures, which are known in other inorganic compounds

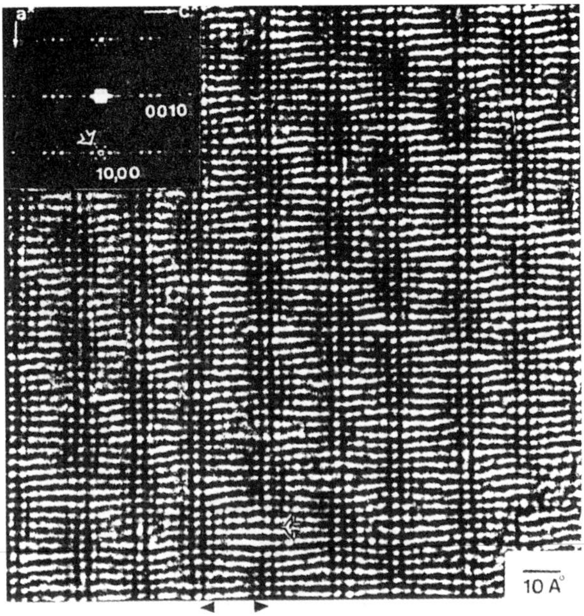

Fig. 52. Electron microscope lattice image of $Bi_2Sr_2CaCu_2O_x$.

(though quite rare), are called 'incommensurate'. Physicists and chemists had a lot of fun working out the structures of the BISSCO family, although in the end the modulation seems to have little to do with the superconducting properties! However, because the BISSCO compounds have such a pronounced layer habit, they form better oriented films than YBCO, and hence have been tried out in several electronic applications (see below).

About a year after the BISSCO phases, T_c went up again with the discovery of closely related compounds containing Tl in place of Bi. The maximum T_c currently known in this series is 128 K but because Tl is extremely toxic it is unlikely that any serious efforts will be made to fabricate devices from it. Finally, a German group reported a T_c of 133 K when the Bi or Tl is replaced with Hg. Again, given the toxicity of Hg, this discovery will most probably remain in the realm of basic science.

Applications of high T_c superconductors

In the years that have followed the discovery of the copper oxide family of high temperature superconductors, many applications have been bruited.

However, following the initial euphoria, quite a lot of obstacles have come up on the way to full technological exploitation, some of which were alluded to earlier. Nevertheless, the first commercially available devices have started to appear. Because it is so hard to make large homogeneous specimens of oxide superconductors, the first applications have been in small-scale electronics, especially in the microwave field. For example, GEC-Marconi put on the market a filter to separate microwave signals of different frequencies based on a one inch square film of the Gd analogue of YBCO deposited on a MgO substrate (Gd was chosen instead of Y because the repeat distance in its structure provided a better match to that of the substrate) (Fig. 53). The device operates at 80 K, and therefore still requires

Fig. 53. High-temperature superconductor planar thin-film microwave filter with a conventional waveguide filter. (Photograph courtesy of GEC-Marconi.)

cooling, but this can be accomplished by a small device that works on the Stirling cycle. Liquid nitrogen is not needed, only electric power. Even including the cooler, such a device is several times smaller and lighter than conventional ones made of copper.

Because miniaturisation is possible, high T_c microwave devices are especially suitable for space applications. For example, trials of microwave antennae for installation in satellites have been made in the USA, with test components having been flown. A less demanding application of superconducting films is in shielding sensitive electronic components from strong magnetic fields. Several Japanese companies have developed chambers lined with film for this purpose, up to 40 cm by 15 cm in size, inside which the magnetic background is reduced by a factor of no less than 10^5. Such boxes must surely constitute the ultimate in Faraday cages!

During the 1970s and 1980s, superconducting tunnel junctions related to the SQUID detectors described above were actively considered by several companies as potential circuit elements in computers because of the very fast switching times that can be achieved. The coming of high T_c materials revived interest in the topic, culminating in an announcement by an American company, CONDUCTUS Inc., that they had fabricated a 32-bit shift register (the most common element in information processing systems) from YBCO. Operating at liquid nitrogen temperatures, this digital electronic circuit runs at a clock speed of no less than 120 GHz, 1,000 times faster than that of a current high-speed personal computer. Incorporation of this device in commercial computers is confidently awaited.

Whilst large-scale uses of high T_c superconductors have not developed so quickly as those in microelectronics, there are one or two indications of viable applications. Current leads for superconducting magnets made of conventional materials are already on the market from Hoechst. The property being exploited here is not so much the superconductivity of the lead, but the low thermal conductivity of high T_c materials, because of their ceramic nature. In a rather more futuristic vein, it has been suggested that disks of high T_c material might form the basis of frictionless magnetic bearings that could carry loads up to 100 kg at speeds up to 500,000 r.p.m. Such flywheels might form the basis of an energy storage system.

The future for superconductors

Only when the ceramic copper oxide superconductors burst upon the world did solid state chemists begin seriously to consider the synthesis of new superconducting compounds as a task falling within their field of competence, rather than that of physicists, metallurgists, and materials scientists. Nevertheless, some years before the *annus mirabilis* of 1987, two quite different series of oxides had been discovered, both of mixed-valence type and both belonging to structural types well known in solid state chemistry, that also behaved as superconductors. As noted earlier, they were respectively $LiTi_2O_4$ (1975), a spinel, and $Ba(Pb,Bi)O_3$ (1975), a perovskite, both with T_c's of 13 K. Since 1987, T_c in the Bi series has gone up to 30 K, with the preparation of $K_xBa_{1-x}BiO_3$. There have also been tantalizing reports of zero resistance in other titanate phases at even higher temperatures, but they remain unconfirmed: chemical stability of defect oxide phases renders such work very difficult.

As long ago as the 1960s the American theorist, Bill Little, promulgated the idea of a new mechanism for superconductivity, which relied on coupling the electrons not to lattice vibrations but electronic excitations. Were substances to be found in which such a recipe could be realised, they would have T_c's measured not in tens, but hundreds, of K. Little's proposal initiated a search especially among molecular-based compounds, the more usual hunting ground for chemists, and finally in 1980 the first molecular superconductor was discovered. The compound in question was a charge transfer salt, that is, one in which a planar organic molecule has lost an electron to form a cation, compensated in the crystal lattice by an equivalent number of negatively charged anions. The first organic superconductors were called Bechgaard salts after the Danish organic chemist who made them. The organic cation was tetramethyl-tetraselenofulvalene (TMTSF for short), and the anions were small inorganic species such as ClO_4^- and PF_6^-. Their T_c's were low (1.3 K), but the important thing is that superconductivity had been carried into a completely new domain of chemistry. Subsequently, many related compounds have been made, with T_c climbing to 13 K. Of course, compounds of this kind are made in quite a different way from ceramics, in fact they are crystallized from an organic solvent electrochemically at room

Fig. 54. Crystal of an organic superconductor mounted for resistance measurements[(15)].

temperature. Nevertheless, wires are attached for resistance measurements in just the same way as to any other metallic conductors (Fig. 54).

The story of new superconducting compounds is far from over: in 1993, one of my former graduate students, Matthew Rosseinsky, working as a postdoctoral fellow at AT&T Bell Laboratories in New Jersey, found that fullerene, the new form of carbon discovered only a year or two before, can also form salts that superconduct. This time they contain inorganic cations—in fact the element discovered by Davy, potassium! The first such salt was K_3C_{60} (T_c of 18 K) but the record in the series is now held by Rb_2CsC_{60}, with a T_c of above 30 K.

Where will it all end? The next practical landmark on the temperature scale must be the temperatures that could be reached by thermoelectric cooling, say some 220 K. That is a long way from the present all-comers' record of 133 K, but remember those old Physics Tripos questions: the mechanism of high T_c superconductivity is still not agreed among theorists sixteen years after its first discovery. So there is no *a priori* reason to rule out even higher

T_c's, and when chemists become involved, the arena of substances to be investigated becomes dramatically enlarged.

On 26 May 1993, the Chancellor of the Duchy of Lancaster, Mr William Waldegrave, in his capacity as Minister for Public Service and Science, introduced the first Government White Paper on Science and Technology for 20 years. In it we find the following statement:

> *Basic research is undertaken to advance fundamental knowledge, irrespective of any foreseeable application. Such speculative research can be a major source of the revolutionary changes in our understanding of the world, producing important technological advances. Many of the dramatic improvements in recent years in our quality of life and standard of living would not have been possible without these discoveries.*

As we survey the history of electrical conductors, from Humphry Davy's 'perfect conductor' through Kammerlingh Onnes's low temperature research to the high T_c superconductors now beginning to find important applications, it would be hard to find a better example than superconductivity to illustrate the point that by finding unusual new properties of matter, new opportunities are created for improving technology and human welfare.

chapter 13

Room at the Bottom

Nano-technology first appeared as a buzzword in the 1980s, when it was taken to mean engineering on a very small scale. 'Nano' means 10^{-9} and 1 nano-metre (nm) is equivalent to 10 Å, about the size of a large molecule on the scale of things. For example, the diameter of a helix of DNA is about 2 nm. In the days before biotechnology became the major growth industry that it is today, the word technology was synonymous with engineering, so the first images of nano-technology were of sub-microscopic machines; of extreme precision in machining surfaces; and of components made to very small tolerances. If only we could make such things, people believed, whole factories would be transformed. The guru of this particular field of futurology was Eric Drexler, who wrote:

> *Assemblers will be able to make virtually anything, from common materials, without labour, replacing smoking factories with systems as clean as forests. They will transform technology and the economy at their roots, opening a new world of possibilities.*

Undoubtedly, very small machines have been made by cutting out components from silicon wafers using electron beam techniques and photolithography. But for chemists, this appears to be a classic engineering approach, starting from above by taking monolithic material such as silicon and carving it into ever smaller pieces. Instead chemists, to whom the world of nano-metric dimensions is entirely familiar, approach the subject from 'below', that is from the scale of individual atoms and molecules. What, therefore, does chemistry have to offer nano-technology? To answer this question we need a more comprehensive view of nano-technology than the one given by Drexler's futurology, and a wider perspective than just constructing tiny mechanical machines. So we need to ask: what kind of requirements is nano-technology fulfilling now and what do we need from it in the future?

In a more general context than Drexler's mechanical one, it could be said that nano-technology draws on all the sciences that manipulate matter on the scale 0.1–100 nm for engineering purposes, just as mechanical engineering uses the ideas and principles of physics, and biotechnology those of biology, to produce manufactured products. Examples already used in products are video recorder spindles with surfaces smooth to within a few nano-metres, and micro-engineered miniature sensors for intravenous medical diagnosis. Viewed against this background, a number of generic themes have begun to emerge:

- modifying and manipulating surfaces;
- elaborating microstructures (usually in two dimensions);
- forming mono-disperse particles;
- synthesising self-assembled cavities and lattices.

Finally, validating and measuring what has been made is fundamentally important so microscopy with nano-metre resolution emerges as a key tool. Let us see where chemistry is contributing, and could contribute, to each of these facets of nano-technology, using some recent examples.

Modifying surfaces at the nano-metre scale requires more than reactive chemical etching or physi-adsorption. It is some years now since precise deposition of films with molecular defined thickness using the Langmuir–Blodgett (LB) method began to impinge on electronic technology. This method exploits the propensity of molecules containing both hydrophilic and hydrophobic groups to form alternate layers by dipping a substrate into a trough of water, which has an amphiphilic substance spread on the surface. The distance between hydrophilic (for example carboxylate) layers is precisely programmed by the number of carbon atoms in the hydrophobic aliphatic chain that separates them. By incorporating functional groups in the organic chains it is possible to induce cross-chain reactions, for example photo-polymerisation, greatly strengthening the film. Useful physical properties, such as ferro-electricity, can be made to emerge by controlling the cooperative reorientation of side-groups in response to external stimuli. My own research group has shown that, quite apart from putting such multi-layers on substrates, there are numerous bulk materials (which we dubbed 'organic-inorganic composites') whose structures consist of similar multi-layers (Fig. 55). However, it is probably fair to say that despite major efforts by chemists and material scientists in the 1980s, neither LB films nor bulk organic-inorganic multi-layers have yet found their way into commercially nano-engineered products. Apart from the extreme clean room conditions needed to make LB films, the main reason why they have not found favour in industry is their poor stability: the film comes off the substrate. Fortunately, in the past few years a solution has been found to this problem by covalently attaching the first molecular layer to the bare substrate. Thus George Whitesides at Harvard has attached alkane-thiols to gold surfaces to give semi-crystalline upright layers of carbon chains bound to the surface by Au-S bonds; with $CH_3(CH_2)_{15}SH$ the films are 2 nm thick. Varying the

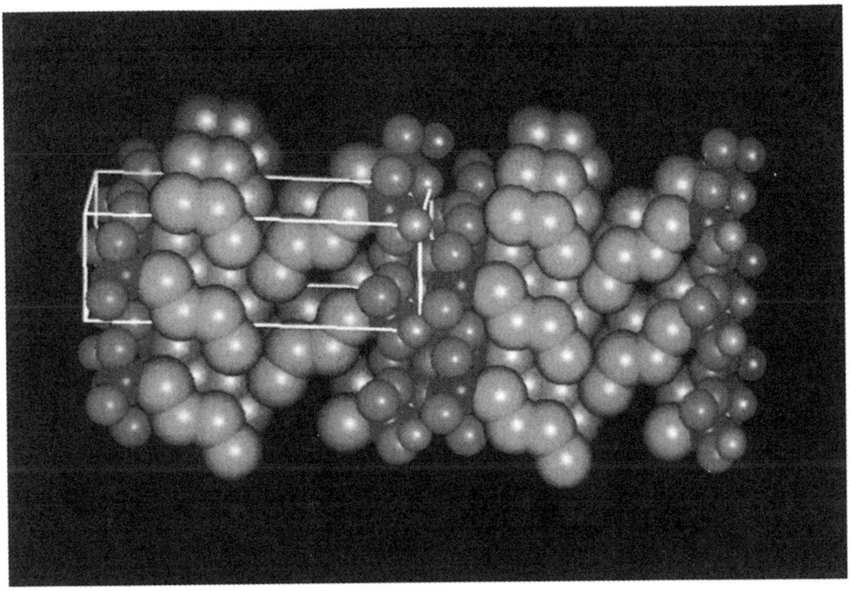

Fig. 55. An organic-inorganic layer composite, $(n\text{-}C_4H_9PO_3)Mn{\cdot}H_2O$.

terminal group on the hydrocarbon chain, by attaching -CF_3 or -COOR for example, yields films with pronounced hydrophobic and hydrophilic character, respectively. Such a mono-layer can then in turn become the substrate for making multi-layers by the LB method, or by direct chemical reaction.

Manipulating surfaces in practice means patterning, something that has been at the heart of the microelectronics industry for 40 years. Most dramatically in recent years, physicists discovered how, using atomic force microscopes, to move single atoms around on clean surfaces and reassemble them into recognisable patterns—pictures of stick men, or 'IBM' written in rare gas atoms. Because such precise positioning can be obtained by purely 'physical' means, it may seem that there is no need for input by chemists. This is not so, for two reasons. First, the engineering requirements of ultra high vacuum and ultra precise positioning etc. are extremely onerous for production purposes: the cost of a fabrication plant for Very Large Scale Integrated (VLSI) circuits now exceeds US1B if we require feature sizes less than 0.5 μm. Secondly, it may be necessary to pattern surfaces that are not completely flat. In the latter context techniques more closely allied to printing than conventional photolithography could be a powerful addition to the microelectronic armoury. For example, the alkyl-thio-gold chemistry mentioned above allows patterning by attaching the other end of the alkyl chain to a master pattern and then transferring it on to the gold substrate, in exactly the same way as inking a printing block.

Nano-metre-thick layers represent one approach, but objects of nanometric size in all three dimensions represent another strong growth area for nano-technology. Nano-particles have been around for many years. At the Royal Institution, we have a sample of colloidal gold made by Michael Faraday in 1856, which still retains its beautiful purple hue. Despite a lot of work by inorganic and organo-metallic chemists over the past 20 years on clusters containing relatively small numbers of metal atoms, only quite recently has it become possible to synthesise macroscopic amounts of well characterised larger clusters. For example Robert Whetton of the Georgia Institute of Technology in Atlanta described a discrete family of gold nanocrystals, passivated by self-assembled mono-layers. Each member of this family has a definite mass and structure: a truncated octahedron of cubic close-packed gold atoms 'coated' by alkyl-thio-groups. One of these clusters contains a staggering 584 gold atoms (Fig. 56). From the point of view

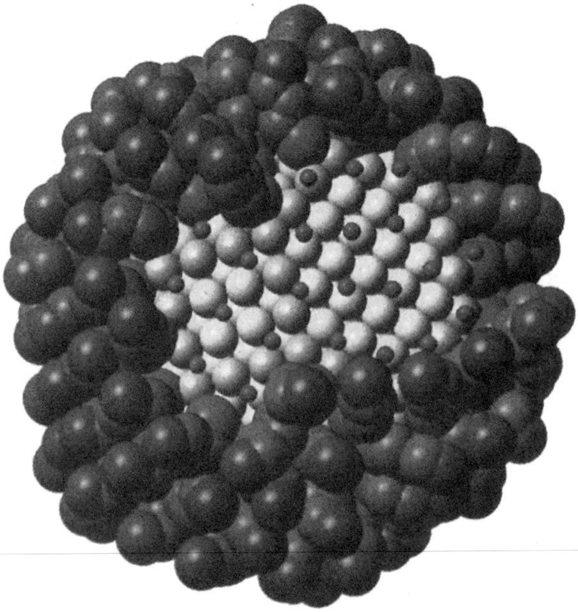

Fig. 56. A cluster of 584 gold atoms.

of its bulk physical properties, a crystal containing such entities has to be viewed as controlled by two very different length scales: the inter-atomic separation, which is essentially the same as in the infinitely extended lattice of gold metal; and the centre-to-centre separation of the clusters. Profound theoretical implications arise from such systems, at the boundary between the discrete and the infinite, of a kind first raised many years ago by physicists Jacques Friedel and Neville Mott.

Possibly the best known nano-metric metal particles are not ones synthesised by chemists, but by biological organisms: the Fe particles in bacteria and pigeons, which provide them with orientational information via the Earth's magnetic field. Chemists would no doubt say that biology is just chemistry writ large, and there is no doubt that the size and morphology of these Fe particles is determined by the growth mechanism, which (though unknown) clearly involves deposition in a nano-structured chemical environment. What is striking about the biologically synthesised particles is that their dimensions are just at the maximum beyond which a bulk piece of Fe would divide into separate magnetic domains—in other words they remain super-paramagnetic. Man-made nano-metric magnetic particles, when suspended in a fluid medium, can impart remarkable bulk properties to the latter and, under the name ferro-fluids, could find applications as packing for rotary transmissions and damping elements in loudspeakers and stepper motors.

Nano-sized 'islands' of semiconductors are already being formed on surfaces by electron beam lithographic methods. The name given to these small groups of atoms by physicists—quantum dots—symbolizes that (as with metals and magnets) the electronic energy levels of semiconductors are strikingly different when only a small number of atoms are involved in the aggregate. In the case of Si and CdS this difference is manifested in the band-gap, so the colour of light emitted is controlled by the particle size. Another way of making nano-metric CdS particles is to form them inside what we might call 'inorganic vesicles'—the cavities occurring in micro- and meso-porous framework structures such as zeolites. Transition metal oxide clusters such as WO_3 placed in the same cavities can be electrochemically oxidised to give coloured material analogous to extended electro-chromic WO_{3-x} thin film displays.

The most famous nano-particles of the past few years are the fullerenes, from C_{60} upwards. Examples are now known with metal atoms inside (e.g. Sc_2C_{62}) though it is fair to say that no clear technology has yet emerged from this fascinating science. When we open the C_{60} cage clusters out into tubes, however, there are even further possibilities. Researchers at the University of Oxford have opened the ends and filled the tubes with metal salt solutions that can be subsequently reduced to metal 'wires'. Attempts are also being made to introduce biologically active molecules into nano-tubes, with the aim of making sub-microscopic sensors.

Finally, among all the points at which chemistry is impinging on nanotechnology, the most tantalising is that of self assembly—a phrase that betokens the way that complex three-dimensional molecular architectures are derived from the shapes and charge distributions of their constituent units. Biology depends on self assembly: the tertiary structure of a protein is inherent in the amino acid sequence alone. In the crystalline phase, great strides have been made over the past few years in designing molecules that come together in pre-defined arrays. This is called crystal engineering, and permits rational design of one-, two- or three-dimensional polymeric lattices. We can discern two principles from self assembly—metal-ligand coordination and hydrogen bonding. An example of the former is the use of a rigid ambidentate ligand such as 4,4′-bipyridine to make tetragonal structures containing large channels of square cross-section, for example $Zn(4,4'\text{-}bipy)_2SiF_6$. Examples of large-scale structures stabilised by hydrogen-bonds abound in nature, in the form of alpha-helices and beta-sheets in proteins and, most famously, the double helix of DNA. Chemists are constructing simple analogues of these—for example the ribbons of pyro-mellitic anhydride and hollow cylinders—in high yields from simple precursors. Such structures do not, in themselves, have chemical or electronic functions but it is not difficult to see that, based on this 'demonstration of principle', nano-scale arrays designed with such functions could soon emerge. Indeed, the tiniest nano-machines are already being made in the form of molecular shuttles, in which a cyclic poly-ether molecule looped around a cyclic bi-pyridinium ion clicks from one stable site to another on a timescale that can be followed by NMR.

These few examples may help to demystify the word nano-technology for chemists; to illustrate some of its diverse facets; and to suggest how chemists are (and will be) playing a vital role in bringing this unusual science to fruition. Unfortunately for chemists though, it was a physicist (and a theorist at that), Richard Feynman, who first coined the phrase that 'there is plenty of room at the bottom'.

chapter 14

Molecular Information Processing: Will It Happen?

Molecular information processing is going on all around us. Information processing based on molecular processes has in fact been going on for millions of years in its natural form, through the brains, not just of human beings and their predecessors, but in all other higher (and perhaps even lower) living organisms. For all brains, however rudimentary in evolutionary terms, operate at the molecular (or more precisely supra-molecular) level. Thus nature has given us an existence theorem and the question becomes a different one: not 'can information be stored and manipulated at the molecular level' but rather 'can we ourselves design and manufacture artificial structures using molecules that will carry out these functions?' Sadly, at the present time the short answer to the second question is 'we don't know', but if that was all there was to be said about the matter there would be no point in continuing this exposition.

So I find myself in the rather unusual position of tackling a subject that does not yet exist. Indeed, in some ways it is even a bit embarrassing because I am not a great fan of crystal ball gazing. Over recent years in the UK the Department of Trade and Industry (the Civil Service resting place for the Office of Science and Technology) has conducted a 'Technology Foresight' exercise, and it is significant that no mention of molecular-based information processing has ever appeared in it. Perhaps this is one result of the very narrow subject base for the advice it has sought: for example no chemist was invited to sit on the panel dealing with information technology (IT) and electronics, despite the fact that even the present generation of integrated circuit hardware is constructed by chemical means using photolithography. Conversely, neither did any representative of the IT industry contribute to the deliberations of the chemicals panel. In fact, when the House of Commons Select Committee on Science and Technology took evidence about the procedure and effectiveness of the Foresight endeavour, I have to confess to submitting (and defending under interrogation) a highly sceptical view on the subject. It is therefore starting from this sceptical and cautious base that I want to consider where the subject that has been called 'molecular electronics' has come from, what its present status is, and what a future information processing regime based on molecules might just possibly look like.

Let me begin by nailing my own colours firmly to the mast: I am a chemist, and a distinguishing feature of chemical science is to build new structures, that is, new arrangements of atoms into molecules and molecules into aggregates (which might be—but do not necessarily have to be—crystalline) in such a way as to create new properties and functions. As noted already, molecular chemistry is already contributing very significantly to the electronics industry as it exists today. Examples are the fabrication of silicon wafers, etching of fine structures on their surfaces using photoresists, deposition of thin films by chemical vapour deposition and, above all, displays made from liquid crystals, which we find in digital watches and laptop computers. However, that is not the focus of my present concern. I want to consider whether molecular assemblies themselves might be used as a means to store and process information. Put baldly like that it sounds like science fiction, but let us remind ourselves what has been happening in electronic data processing over the last 50 years, or indeed in other forms of processing for longer than that.

The earliest mechanical form of data processing (if you leave aside counting on our fingers, which gives rise to the word 'digital') was the abacus. The simple form of this device, still widely used in bazaars throughout the Middle and Far East, serves to illustrate a number of important points about computational functions that are shared by advanced electronic systems. First of all the information is stored and processed in 'bits', units symbolized by the balls that move on the wires of the frame. Second, the bits are manipulated serially, i.e. one after the other, rather than all at once. Third (a point that will come up again later), to move the bits takes a certain amount of energy, in this case muscle power, thus reinforcing the fact that each computational step consumes power. The same considerations apply to more advanced mechanical computers, which reached their apotheosis in Babbage's arithmetical engine, a device of gear wheels so elaborate that the handle could scarcely be cranked by a strong man. In more recent times mechanical calculators remained in use till some 30 years ago, overlapping chronologically with the first ones to use electrons.

The earliest devices to process information using electrons, such as the Colossus at Bletchley Park during the Second World War, were no smaller than Babbage's engine: indeed, the Colossus (which was well named) filled a whole room. Similarly, the individual components (valves) from which they were constructed were of similar dimensions to Babbage's gear wheels, i.e. a few centimetres. Furthermore, the programming and recording mechanisms remained mechanical, in the form of holes punched in cards or paper tape. Over the succeeding 50 years, however, the components of information processors have grown inexorably smaller, at the rate of one order of magnitude every 10 years. Valves were replaced by transistors, so that the electrons no longer circulated in a vacuum but through the conduction band formed from the overlapping atomic orbitals in a solid. Single transistors, linked by wires, gave way to integrated and then to very large scale integrated (VLSI) circuits etched by elaborate sequences of chemical operations on to the surface of a silicon single crystal. Figure 57 gives a pictorial indication of the orders of magnitude that have been traversed.

Nowadays, the individual features found in commercially available personal computers are less than 1 micron (one-millionth of a metre) across and compare with the size of a virus. Enormously complicated patterns are achieved routinely by photolithography (Fig. 58) or etching with finely focused electron beams. Astonishingly, this evolution in the size of electronic computing elements, or features, follows an exact linear relationship with time, over quite a long period (Fig. 59). It has been called Moore's law, after the founder of the integrated circuit manufacturing company, Intel. In fact the relation shown in Fig. 3 turns out not to be the result of some hidden law of nature, but actually results from economics: the cost of

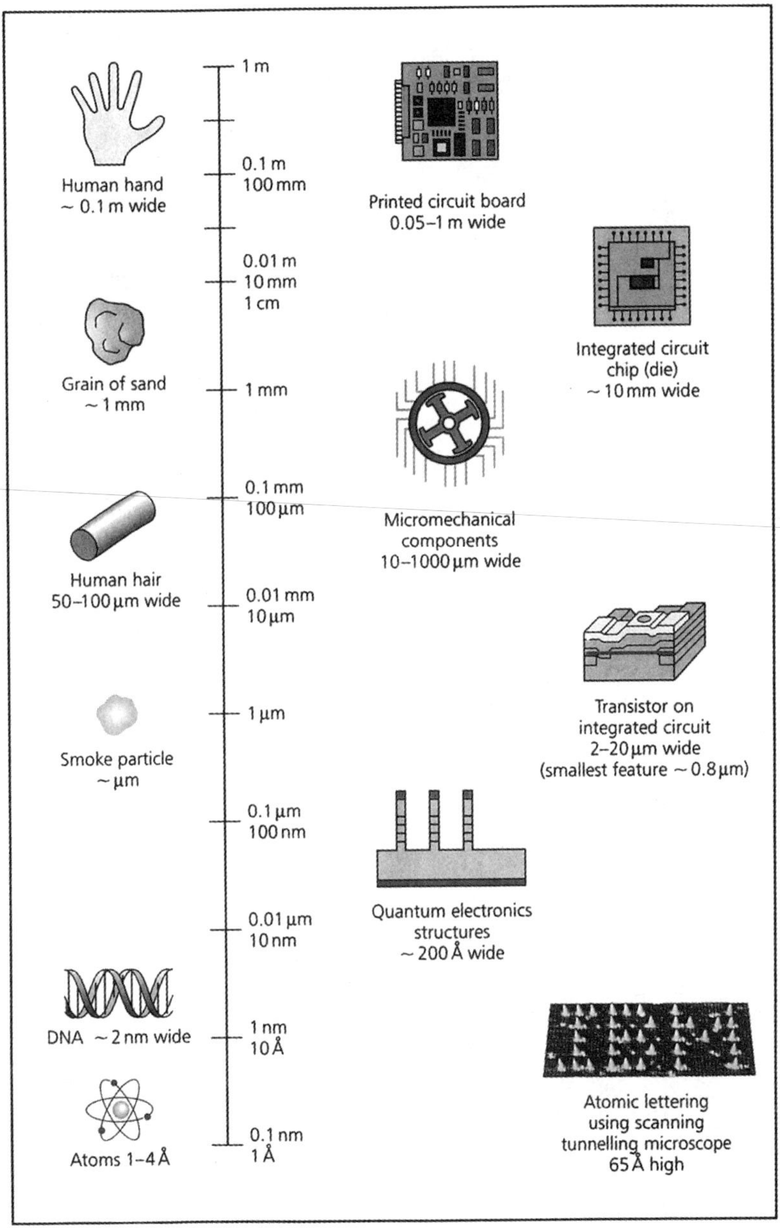

Fig. 57. A pictorial representation of the sizes of information storage media. Reproduced with permission from *Chem. Brit.* **32**, 29–31 (1996).

Fig. 58. A modern microprocessor. Reproduced courtesy of Sharp Electronics.

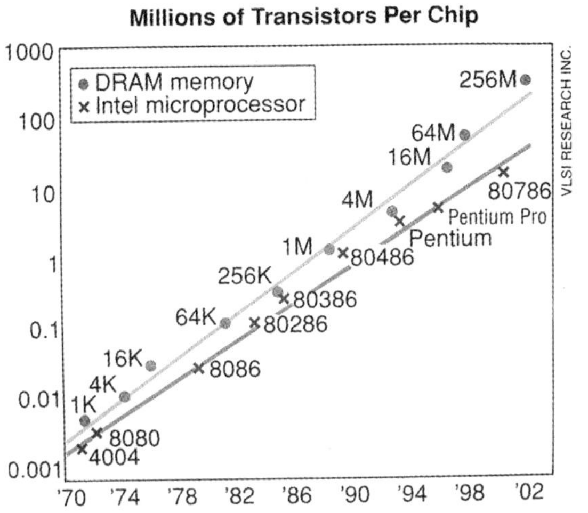

Fig. 59. Decreasing feature sizes of microprocessors with time. Reproduced by permission from *Science* **274**, 1834 (1996).

building a new plant to produce microprocessors scales with their feature size, and currently it stands in the region of US$1B, which explains why chip manufacturing throughout the world is concentrated in so few hands.

Given the vast size of the investment in time, effort, ingenuity (and hence money) to make microprocessor features ever smaller and smaller, it is legitimate to ask why all this trouble is being taken. Clearly, small size is an advantage in itself: we can carry around computing power in a laptop equivalent to the mainframe machines of 20 years ago. Individual switching processes also take less power. Valve-based computers had huge power consumption but the laptop only needs a rechargeable battery. (Incidentally the latter is itself a marvel of solid state chemistry, but that is a story for another time.) In view of these advantages to miniaturization, can we imagine any limit to the process, apart from human ingenuity and cost? Well, unfortunately, yes. There are three so-called 'fundamental' limits, set by the laws of nature. The first is set by the statistics of the switching process in a binary system (i.e. on or off), which translates into thermodynamics via entropy. The second arises from the dictates of quantum mechanics, embodied in Heisenberg's uncertainty principle. The third, which is the most important one in practice, is determined by the heat dissipated by each switching event. As we saw with the abacus and Babbage's engine, it takes energy to throw a switch, and it is no accident that the largest volume inside the central processor unit of a modern supercomputer consists of pipes filled with refrigerating fluid! Altogether the three limits can be displayed on a single plot of power dissipation against switching speed (Fig. 60).

The final consideration brings me at last to the central point. Single chemical events, such as the reaction of a molecule catalysed by an enzyme, consume far less energy than the present generation of semiconductor-based electronic switches (roughly 10^2 kT compared with 10^{10} kT). In principle, therefore, substantial advantages might follow from exploiting molecular chemical events as elementary information processing. If that is the case, several questions need to be addressed before we could contemplate any attempt to realise the potential of the new approach. First, is it

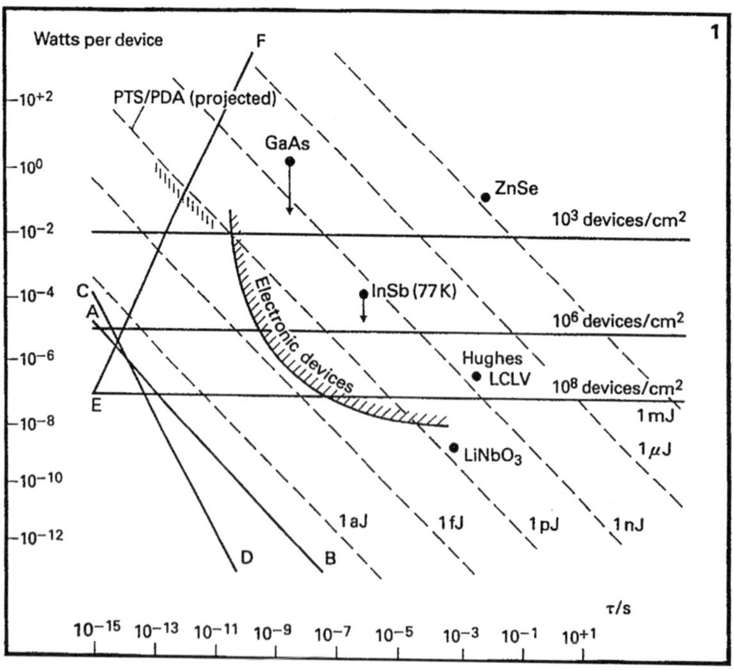

Fig. 60. The fundamental limits of computing speed and power dissipation. Reproduced by permission from *Chem. Brit.* **26**, 52–54 (1990).

going to be possible to make complex ordered structures from molecular components analogous to those of microelectronic circuits such as the ones in Fig. 58? Second, what are the kinds of physical or chemical processes that could be envisaged to store and process the information? And third (perhaps the hardest question), how could we get the signal in and out? Or, put another way, can we interrogate and detect the states of single molecules? The remainder of this chapter will be devoted to providing some answers to these questions.

Building structures out of atomic or molecular building blocks that are ordered over a long range in one, two, or three dimensions is at the very heart of the science of solid-state chemistry. For example, arrays of molecules forming domains with different orientations are shown in Fig. 61, where the components in question are long chains of carbon atoms (n-alkanes). Note the great, although superficial, similarity between this structure and that of Fig. 58. One of the most powerful methods of engineering ordered molecular arrays exploits the affinity of hydrophobic molecules,

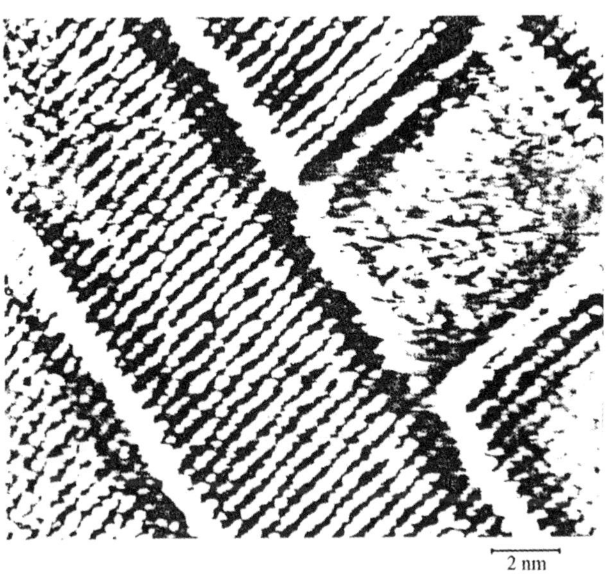

Fig. 61. Structural domains in an n-alkane crystal. Reproduced by permission from *Atomic and Nanoscale Modification of Materials*, P. Avonis (ed.), Kluwer Academic Publishers, Dordrecht, 1993, p. 263.

such as the alkanes above, for other hydrophobic ones, and of hydrophilic (literally 'water loving') for hydrophilic. It is termed the Langmuir–Blodgett method, but I will not give more than the briefest of mentions here. This approach to making thin films one molecule thick on the surface of water is said to have had its origin in an experiment carried out by Benjamin Franklin, who poured a small quantity of oil on to the Serpentine pond in Hyde Park and, watching it spread out across the entire surface, was able to make a simple calculation of its thickness from the volume of the oil that he had poured out.

Apart from such rough and ready arithmetic, there are nowadays many ways of verifying that such films really are only one molecule thick. For example, one can build up multi-layers by dipping a glass slide repeatedly in and out of a trough of water, which has such a monomolecular film on its surface. If the molecular chains are terminated by hydrophilic groups, they stick in the water, and are transferred to the similarly hydrophilic silicate surface of the glass. Thereafter, successive dippings produce layers on the glass with successive hydrophilic–hydrophilic and hydrophobic–hydrophobic orientations (Fig. 62). How do we know that single layers are being added at each stage? If the molecules each contain a light absorbing centre we can measure the absorption increase after each dipping: the increment turns out to be constant. Even better, atomic scale resolution in electron microscopy permits a direct view of the aligned molecules, as seen in Fig. 63.

Apart from these somewhat artificial methods of producing thin films of molecules, three-dimensional architectural assemblies of molecules are

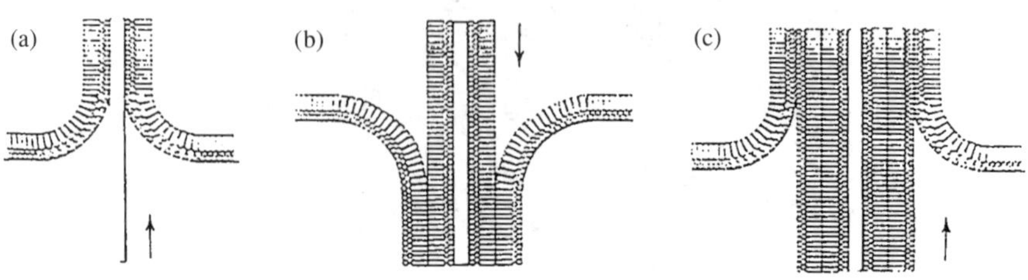

Fig. 62. Forming Langmuir–Blodgett films. Reproduced by permission from *Introduction to Molecular Electronics*, M. C. Petty, M. R. Bryce, and D. Bloor (eds.), London, Edward Arnold, 1995, p. 225.

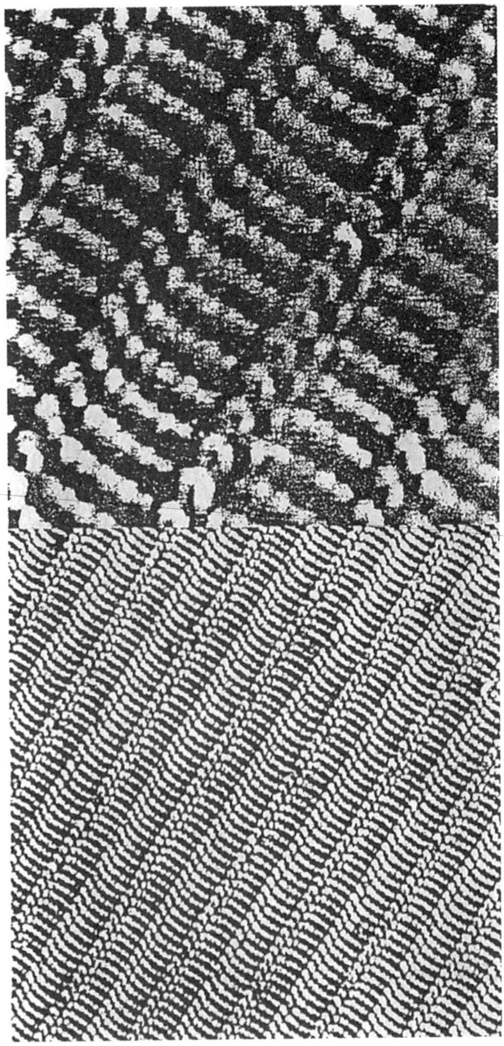

Fig. 63. Aligned molecules observed by electron microscopy. Reproduced by permission from *Adv. Mater.* **8**, 903 (1996).

produced in wondrous variety by chemical synthesis. Figure 64 brings together a few examples culled almost at random from the literature of the last few years. Such elaborate constructions are not put together by what have (somewhat disparagingly) been called 'engineering methods',

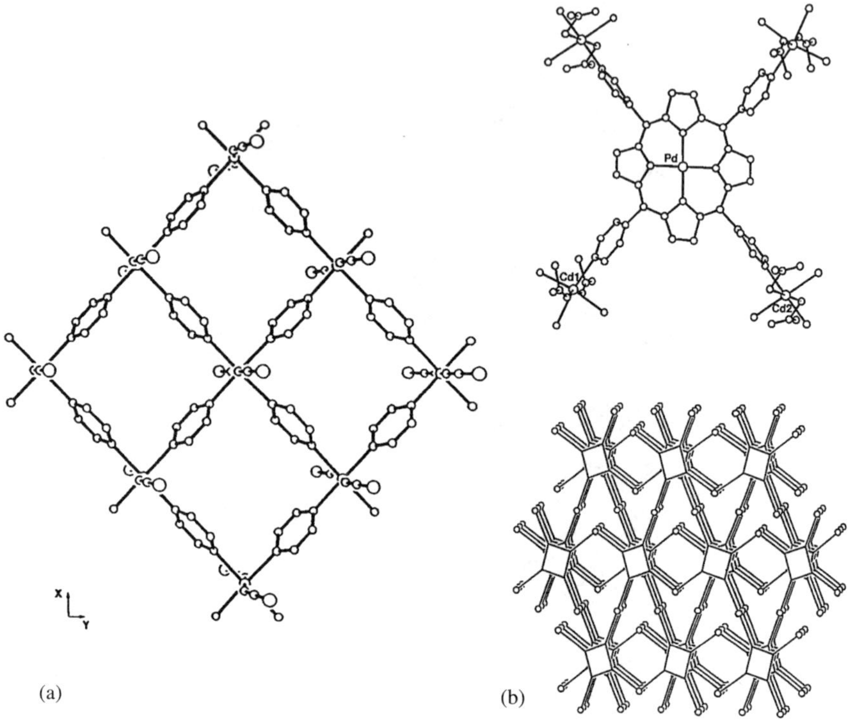

Fig. 64. Examples of self-assembled molecular arrays. (a) Iron pyrazine thiocyanate: (b) palladium porphyrin cadmium adduct.

i.e. piece by piece assembly, but by the much more powerful method of ‘self-assembly’. The latter reaches its apotheosis in biology, where the three-dimensional structures of enzymes are completely defined by the sequence of amino-acids forming the protein chain (the so-called primary structure). Hydrogen bonding coils the primary structure into the alpha-helix (secondary structure), which in turn coils into the final shape through the medium of charge, and both hydrophobic and hydrophilic interactions.

An example of a low-dimensional magnetic lattice, which self-assembles merely by mixing the ingredients in aqueous solution, is the phosphonate salt synthesised by Simon Carling at the Royal Institution (Fig. 55). The organic and inorganic components in the structure segregate spontaneously into alternating layers. Further examples from our own work in the

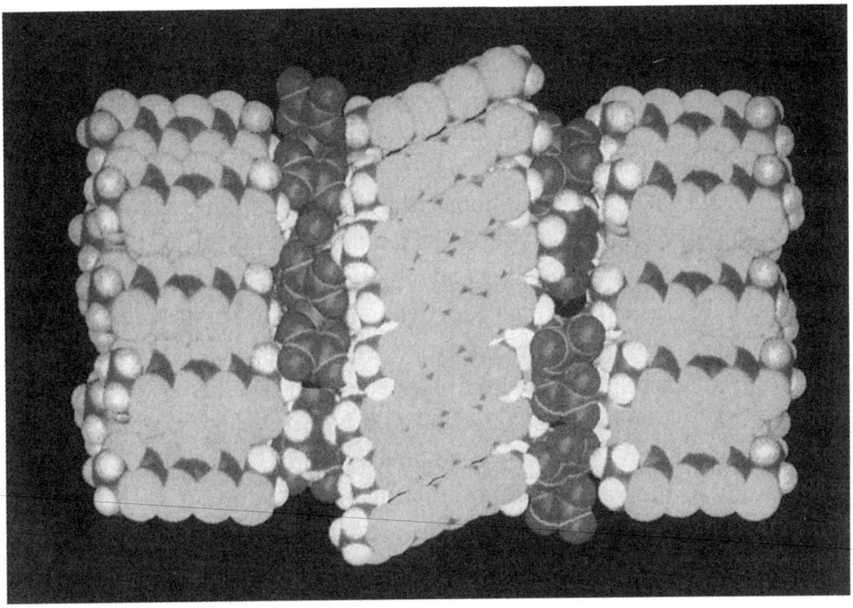

Fig. 65. Organic and inorganic layers in a superconducting molecular charge transfer salt synthesised in the Davy Faraday Research Laboratory of the Royal Institution.

Davy Faraday Research Laboratory are molecular layer compounds showing the highly unusual property of superconductivity (Fig. 65).

The compounds in question are so-called 'charge transfer salts', formed by oxidising a heterocyclic aromatic molecule called bis(ethylenedithio) tetrathiafulvalene (or ET for short) and combining the resulting cations with inorganic anions. The oxidation is carried out most conveniently by electrochemical means and indeed, it has often occurred to me that the small glass cells used for growing crystals of these materials would be recognised quite easily by Sir Humphry Davy, should his ghost return to our laboratory!

The structures of one of the compounds prepared in this way is shown in Fig. 65. Again we see the organic part of the structure (ET) segregated from the inorganic part, $[(H_3O)Fe(C_2O_4)_3{\cdot}C_6H_5CN]$. The compound is metallic, and becomes superconducting at low temperature. It is worth noticing in passing that we can now make superconductors containing water inside the crystal lattice, something that would have been thought quite extraordinary a few years ago. The compound in Fig. 65 also contains a high concentration of paramagnetic iron atoms, again something that would have been thought quite incompatible with superconductivity before, because magnetic moments are supposed to disrupt the Cooper pairs carrying the superconducting current.

Apart from self-assembly of molecular units into infinite arrays and networks, another development of recent years that has given a big impetus to the new field of nano-technology and molecular electronics is the controlled preparation of clusters of atoms with defined sizes, and hence properties. So far as small metal particles are concerned, the origins of this topic lie far back in history, and actually have their origins in the Royal Institution. Michael Faraday made a colloidal solution of gold whose beautiful purple colour remains perfectly clear and un-coagulated to this day: it rests in a cupboard in the study of the Royal Institution's Director. Much more recently it has become possible to synthesise particles containing quite distinct numbers of atoms; for example, one contains 584 gold atoms (Fig. 56). That seems a strange number, but you have to bear in mind that the packing of the atoms is the same as in bulk gold, that is, a hexagonal close-packed lattice of equal spheres. Starting from a single central atom, we can imagine it surrounded by 12 neighbours, then a further layer is added, and so on, generating a series of 'magic numbers' corresponding to the number of shells added. Finally, the particle is sheathed in an organic coating of alkanethiol molecules, whose sulphur atoms attach themselves to the outermost gold atoms. Similar particles can also be made from semiconductors such as CdS and in that case the luminescence that takes place when electrons are excited across the band gap varies very markedly in frequency with the size of the particle. Particles about 20 A in diameter emit blue light while those 40 A emit red. Sizes in between emit every other colour in the visible spectrum. These are examples of what physicists now call 'quantum dots': electrons being confined in such small volumes that their energies are quantised.

Even more remarkable than the ability to manipulate the sizes of such tiny objects is the way in which it is now possible to move them around on a surface. Even atomic scale patterns have been made using the atomic force microscope: a famous early instance was writing the name of the company IBM in xenon atoms on a silicon surface. A circle of iron atoms on such a surface also acts as a 'corral' for electrons trapped inside it, whose wave-like energy states are reminiscent of the waves seen on the surface of the tea in a cup when it is shaken. Unfortunately, in many cases the atoms do not form very strong bonds with the surface, so they have to be placed and maintained at low temperatures. A significant exception, however, is the icosohedral carbon cluster molecule buckminsterfullerene (C_{60}): it turns out that they stick hard enough that they can be manipulated at room temperature, as, for example, into the S-shaped array illustrated in Fig. 66. From here it is just a short step back to the abacus mentioned earlier, and C_{60} molecules have actually been moved back and forth in a groove on a silicon surface to make a nano-scale model of this ancient counting device.

If the answer to the question 'Can we make ordered atomic and molecular arrays with interesting electronic behaviour by chemical methods?' is a resounding 'yes', then what about the second question about the mechanisms that could be envisaged to store and process information? It was in this area that much of the early hype about molecular electronics gave the subject a bad name and led to justifiable scepticism. Over 20 years ago an American theoretical physicist, Forrest Carter, began to speculate about the construction of gates and switches (necessary building blocks for information processing) out of molecular components. Many possible switching mechanisms were taken into account such as jumping of protons from one side to the other of unsymmetrical hydrogen bonds, electron transfer between donor and acceptor groups, and so on. Carter's ideas created a big stir and several conferences were held to discuss them. However, being a physicist (and a theorist at that) Carter had not really given much close thought to how such elaborate molecular structures as he conceived might be synthesised. His ideas therefore met with much scepticism, especially among chemists. An example of one of his proposals is shown in Fig. 67, which shows a hypothetical molecular wire with a gate attached to it.

Now in fact there is some practical basis behind the idea of a molecular wire, as such wires exist in nature in the form of conjugated polymers, of which the simplest prototype is poly-acetylene, a chain of carbon atoms connected by alternating single and double bonds. If we represent it in the *trans* configuration, it is clear that we can write two equivalent structures,

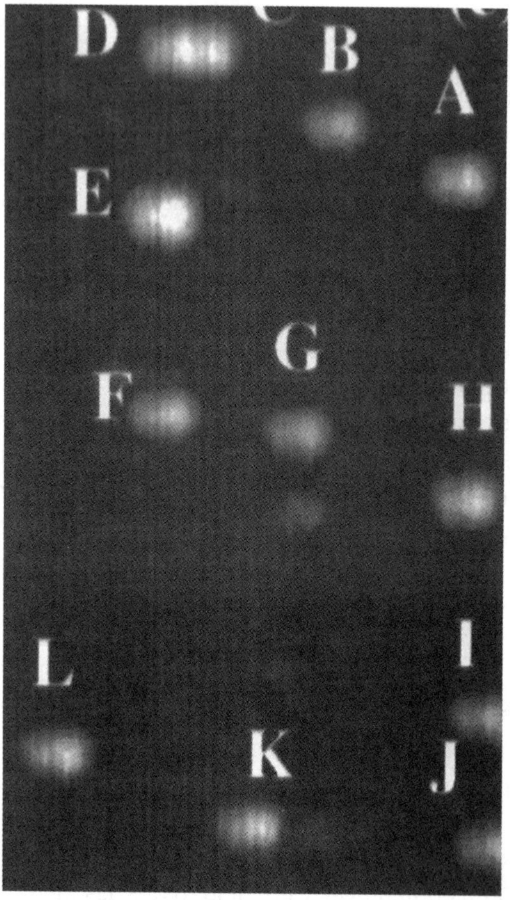

Fig. 66. Fullerene molecules placed in an array. Reproduced by permission from EPSRC Newsline, November 1996, p. 7.

in a fashion analogous to benzene:

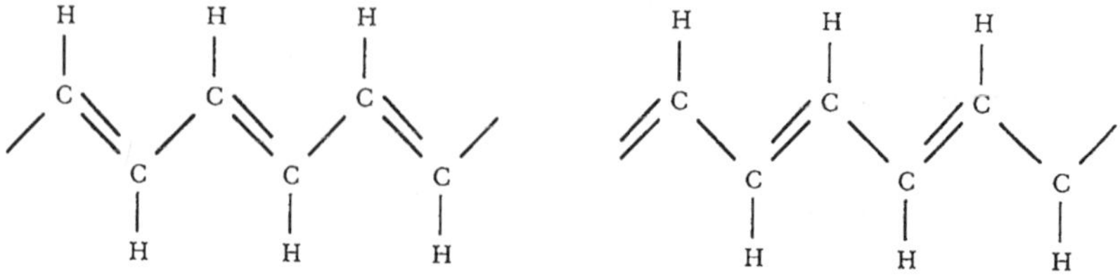

In an infinitely long chain the two would be indistinguishable but consider what happens if we reverse the conjugation in the middle of a chain:

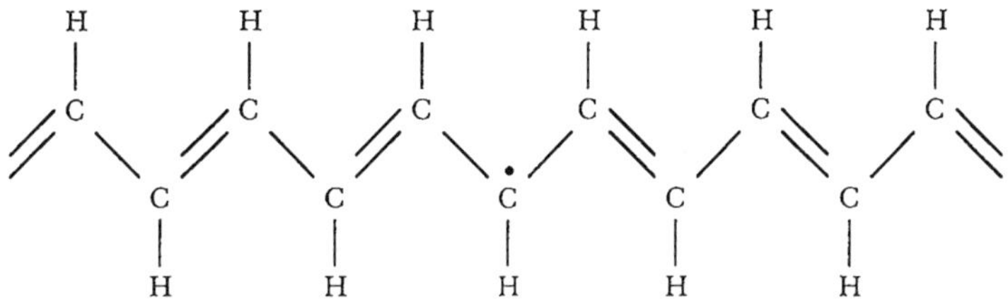

The dot signifies the presence of an unpaired electron separating the two parts of the chain where the alternation has different phases. By exchanging single and double bonds this singularity can move along the chain, taking the unpaired electron with it. It turns out the motion is analogous to that of a solitary wave, of which the best known example is the Severn Bore, the single tidal wave that sweeps up the River Severn every 24 hours. In the context of wave-particle duality, the solitary wave can be identified with a pseudo-particle called a soliton.

Apart from solitons, several other developments in molecular science in the last few years lend hope of purely molecular data processing events. The heart of a transistor is a p-n junction, a region of semiconductor separating positively and negatively doped material. Such junctions act as rectifiers in the sense that on applying a potential in one direction current can be made to flow, but not in the reverse direction. A hypothetical molecular switch was the subject of a conceptual patent by Aviram and Ratner as long ago as 1974, but recently British scientists Ashwell and Sambles have

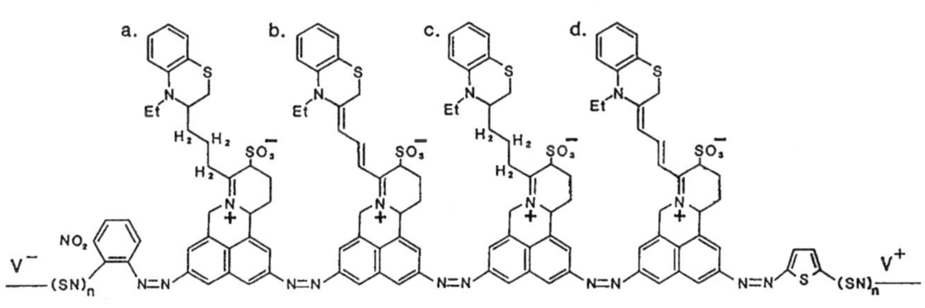

Fig. 67. A hypothetical molecular circuit gate (after Forrest Carter, 1984).

realised the effect by building molecules that combine electron donating and accepting components into Langmuir–Blodgett films. A second, and quite different, approach has been taken by Stoddart, who has synthesised what he calls 'molecular shuttles'. A cyclic molecule forms a loop around a linear one or another much bigger cyclic one, and the loop moves from unit to unit up and down the chain. A method called dynamic proton nuclear magnetic resonance shows that in an example such as that of Fig. 68 the 'train' moves on the 'Circle Line' at about 300 turns per second. Two trains can be introduced using high pressure synthesis, when it turns out that they rotate at the same speed, never colliding! Other possibilities could involve the motion of ions, which is what happens when nerve impulses travel, or the transport of chemicals, which in biology is accomplished by neurotransmitters. All of these are chemical options for transmitting impulses on a molecular dimension, but for all of them there remains a fundamental question, the third on my list.

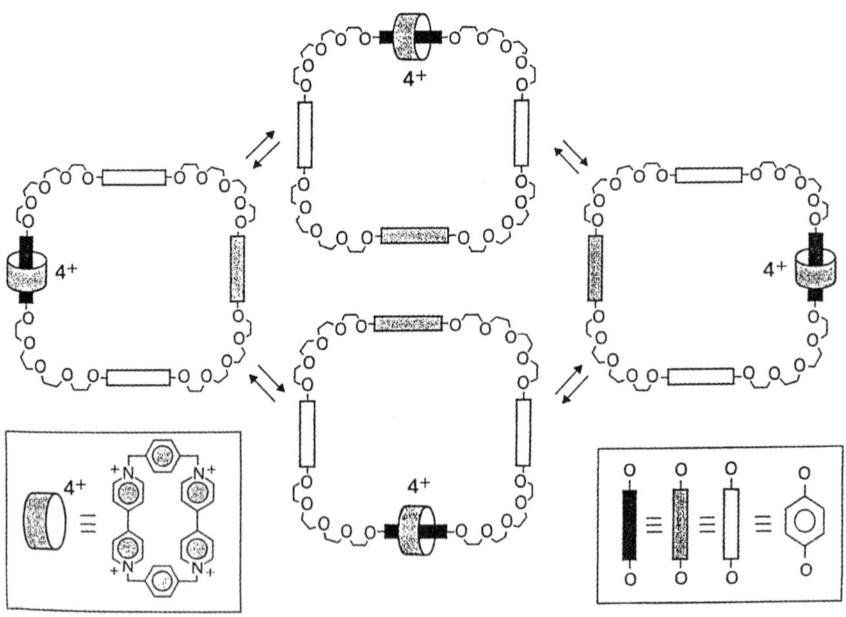

Fig. 68. A molecular shuttle.

If the ultimate goal is to carry out data storage or processing at the truly molecular level, we have to face the question 'how do we get the signal in and out?' or, more starkly: 'can we envisage ways of addressing the status of a single molecule?' At this juncture we have arrived at the point where analogies with conventional semiconductor electronics break down, despite the fact that (as we have already seen) it is already perfectly within the realm of possibility to manipulate molecules, or even single atoms by atomic force microscopy. It is also the point at which the conceptual models like Forrest Carter's fail, even if one could ever persuade a team of synthetic chemists to try and make one. The problem arises from a simple failure of imagination: why should the architecture of a molecular information processor mimic that of one made from silicon?

Let us consider what tasks an elementary information processing unit has to perform, in an entirely abstract way. Such tasks are often set out in the form of a so-called 'truth table', summarising the relation between the signal going in and the one coming out. Usually, the signal is assumed to be binary (i.e. either 0 or 1), but this is not essential. Two of the simplest of such gates, shown below, mimic the functions AND and OR:

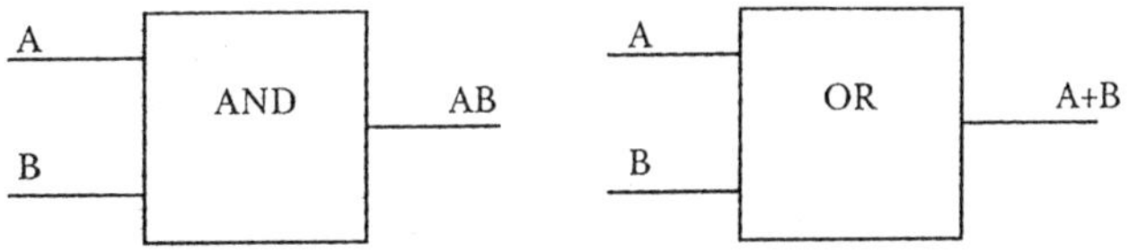

Now the rectangles are 'black boxes': what goes on inside them does not matter. From our point of view, are they single molecules (*à la* Forrest Carter) or something more complicated?

In conventional microelectronics the signal, consisting of a bunch of electrons, passes around the circuit from one element to another. Another way of expressing this is to say that the processing is serial. Now consider a set of individual elements that interact, not in a one-dimensional (serial) way but in two or three dimensions. In that case the state of any one element is determined by the number of its neighbours in the same state. Thus is born the notion of co-operativity, leading to bi-stable states when the whole set undergoes a transition from one state to another. Such phenomena have been used to store data on a much larger length scale for many years, for example in magnetic memory devices. The ultimate requirement is for the individual elements to be capable of existing in two states (charge, spin, vibrational; it doesn't matter), with a mechanism for each element to be sensitive to the state of its neighbours. One practical example of this kind, where the elements are transition metal ions bridged by organic molecules, already exists in the form of lattices in which metal ions change their spin states. Indeed, it is entering service with France Telecom as a means of reading phone cards.

Even in the realm of conventional microelectronics, structures are coming to resemble more closely the requirements of a molecular array. As the 'feature size' (jargon for the individual junctions and processing elements) becomes smaller and smaller, one arrives, willy nilly, at a situation where they all interact with one another, so that it is no longer possible to define a unique pathway carrying the electrons across the surface of the chip. Furthermore, to optimise the manufacturing yield and minimise the number of lithographic errors, designers have to make the units simpler and their arrangement more regular. Such a regular connected array, each node of which can exist in either of two states (on, off, or 0, 1) is an example of what has been called a 'two-state cellular automaton' (Fig. 69). The point

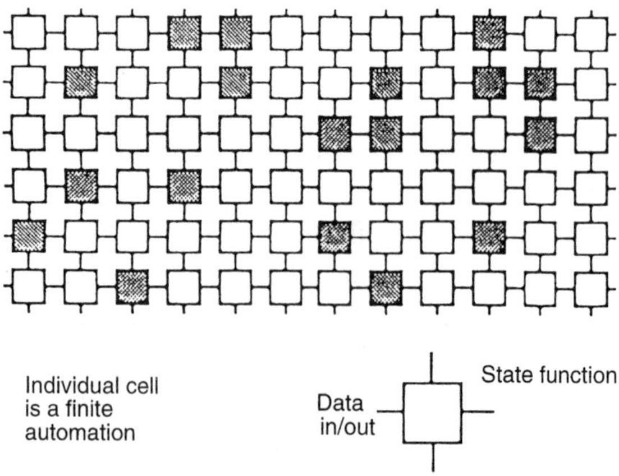

Fig. 69. A lattice behaving as a two-state automaton. Reproduced by permission from *Introduction to Molecular Electronics*, M. Petty, M. R. Bryce, and D. Bloor, Edward Arnold, London, 1995, p. 357.

here is that the entire lattice behaves as a single system. To store or process information you do not have to interrogate each element to find out its state: the information is stored in the form of a pattern. This idea forms the basis of Conway's famous Game of Life, a program available for most personal computers.

Having set up the lattice on the screen one defines an algorithm which switches the state of a given element when a given number of neighbouring elements are in the same state. Starting with a group of elements in the 'on' state (symbolising for us the injection of a signal, for instance at the edge of the array) the pattern evolves from generation to generation, perhaps exiting on the other side of the array. Figure 70 shows an example. This idea is not new; it has been investigated theoretically by Hopfield and Wolfram in the

Fig. 70. Evolution of patterns in a two-state automaton. Reproduced by permission from *Introduction to Molecular Electronics*, M. Petty, M. R. Bryce, and D. Bloor, Edward Arnold, London, 1995, p. 357.

USA, and in the UK by Barker. What has it got to do with molecular-based information processing? Well, the smallest kind of regular array of objects is a crystalline one formed from molecules, each of which might have a different charge, spin, vibration, conformation, protonation, or any other kind of state we might dream up. The beauty of molecular arrays is that, as we have seen, for example in Fig. 64, given a particular shape and charge distribution among the constituents, they assemble themselves, in contrast to the need for lithographic writing to create patterns on silicon. But now we are straying towards science fiction, and it is the moment to sum up.

As a platform for thinking (and perhaps also action) towards an artificial information storage and processing regime based on molecular ingredients, let us consider a number of clear facts. First, since the dawn of computation, the individual information storage and processing entities found in computing machines have been inexorably getting smaller and smaller. Second, chemistry is good at making precisely defined and structured arrays of molecules, capable of existing in different electronic, magnetic or whatever kinds of distinguishable states. Third, methods do exist whereby single molecules can be addressed, although fourth, and more subtly, perhaps it is not necessary to do so. Does all that make it likely that we shall see molecular-based information processing in practice one day? Finally, of course, we do not know but, for myself, if I could be here in 50 years time to collect, I would bet on it. After all, our brains make a pretty good job of it already!

chapter 15

Connecting Atoms with Words

If we agree with the Bible that 'of the making of books there is no end', then it follows that the reviewing of these books for the interested reader is likewise an unending endeavour. Over the years, sundry organs (to use the portmanteau word beloved of *Private Eye*) have been kind enough to solicit my opinion about books on inorganic and molecular solids and their properties on behalf of their readers. Below are reproduced a few of these that seem to me, with the passage of time, to have had something useful to say about the present state and future prospects of the subject to which I have devoted much of my own intellectual energy. The views expressed are not only, therefore, about the books themselves but try to convey some spirit of the way that this part of science was evolving (or, in my view, ought to have been evolving) at the time they were written. Perhaps then, they may contain more than an opinion about an author's work.

Low-Dimensional Materials

Low-dimensional compounds are ones having very an-isotropic physical properties: cooperative magnetic interaction or electron transport is much more pronounced within chains or layers of closely spaced atoms than from one to another. Many organic and inorganic compounds are low-dimensional in this sense (graphite is a simple example) but the main reason why they have become so interesting is the discovery of the one-dimensional 'molecular metals' and the realisation that these compounds, as well as their magnetic analogues, offered unprecedented opportunities to verify some of the most fundamental theorems of solid state physics. Add to this the chance of discovering the elusive high temperature excitonic superconductor, and both University and industrial materials research workers (U.K. scientists conspicuously excepted) flocked into the field. Much of the fruit of the first decade's activity in this topic was distilled into the proceedings of a symposium held at the New York Academy of Sciences in June 1977. One must report, regretfully, that no excitonic superconductor has yet been found, but the range of substances reported on (covalent polymers, organic metals, mixed valence chains) is striking evidence of the vitality of this new kind of materials science. Most valuable of all as a pointer to the future was, in my opinion, the destruction of old boundaries between physics and chemistry. Here was a new breed of synthetic organic chemist who knows what a Brillouin zone is, and of solid state physicist who is not frightened by coordination complexes and organo-metallic sandwich compounds. An exciting new discipline came of age.

Synthesis and Properties of Low-Dimensional Materials, ed. J. S. Miller and A. J. Epstein, N.Y. Acad. Sci., 1979.

Linking Molecules into Solids

During the 1950s it became commonplace among chemists to refer to the 'renaissance of inorganic chemistry'. Of course, as the subject that takes its definition from the preparation and properties of compounds covering the whole Periodic Table, it is hard to imagine that it could ever have suffered a dark age from which to revive. Yet so it was.

Constructing new and elaborate architectures of atoms had came to be thought of as the province of organic chemists; understanding their shapes and chemical reactions the department of the physical chemists. Inorganic compounds were rather boring oxides and halides or salts of simple acids like nitrates and sulphates. The world of coordination complexes, in which a central atom was surrounded by, and bound to, other atoms or groups of atoms (called ligands) had not been seriously transformed since before the First World War, when Alfred Werner had described these structures in work that gained him a Nobel Prize.

What happened to organic chemistry in the 1950s was that the advent of a generation of skilled workers in the synthesis of metal-organic complexes coincided with the popularising of a theoretical model called crystal or ligand field theory. This explained many of the properties of the compounds nicely by describing the central metal in terms of how the electrical fields of the surrounding atoms affected it. You will notice that I did not write 'all the properties'. This is because the model was much better at static properties like stereochemistry (the arrangement of atoms in space) and certain electronic properties like the visible spectra of transition metal compounds. Indeed, ligand field theory had nothing at all to say about the compounds of these two-thirds of the Periodic Table's elements that are outside the so-called d-block. One of the encouraging things about Duffy's book 'Bonding, Energy Levels and Bands in Inorganic Chemistry' is the frequent mention of compounds containing no partly filled d-shells.

The renaissance of inorganic chemistry in the 1950s was narrowly based: it was concentrated on transition metal compounds, and especially on ones that were molecular in type, whether uncharged molecules soluble in organic solvents like most organo-metallics, or more elaborate versions of the classical Werner-type coordination complexes. In particular, though, the ligand field model was only capable of describing assemblies of such molecules (crystals, for example) as the sum of their constituent units. Any properties, such as electrical conductivity, or optical non-linearity, that can only be defined in terms of the aggregate bulk material, are quite outside its scope. For example, the idea that a molecule is a conductor is meaningless: only a lump of matter of macroscopic size (say one or two millimetres on edge) can be subjected to such a measurement by attaching pieces of wire to it and watching if a current flows. That is the paradigm of the dilemma that forms the point of departure for another renaissance of inorganic chemistry and subject matter of Duffy's book.

While the inorganic chemists were worrying about small shifts in the absorption spectra of coordination compounds when pyridine replaced ammonia as a ligand, the physicists, in their beady-eyed way, were sorting out the categories of solids and asking simple questions such as why some crystals conduct electricity and others do not, and why (in the case of nickel oxide for instance) the conductivity can change by ten orders of magnitude if the oxygen content changes. Where the theoretical physicists Mott and Anderson led, the chemists followed finally. They discovered that they could even outdo the physicists by presenting them, not only with more good examples of their rather abstract models but (much more entertainingly) new arrangements of atoms, to which the simplistic models of the physicists, reared on elements and binary compounds, such as sodium chloride or zinc oxide, did not apply.

So the ball came back to the chemists' court. Still, they will only pick it up and run with it if they understand the principles that the physicists enunciated. It is in this context that we must approach Duffy's book. There are not many books that aim to introduce undergraduate chemists to the physicists' models, shorn of mathematical detail but with enough examples of substances and their properties to whet a chemist's appetite for the immediate and concrete: What composition? What colour? Metal? Magnet?

Another interrogative follows those: does the book succeed? With sadness, not entirely. It rests too much in the past of the 1950 renaissance of inorganic chemistry. As a starting point for introducing a chemical audience to solids, atomic orbitals sound right, but it does not have the detail reminiscent of a 1960s book on ligand field theory. More damaging is the small range of chemical examples. Quite fairly the emphasis is on optical properties, since absorption and emission of photons provides marvellous tools to map electronic structure. Oxides, too, were the key examples that gave Mott the electronic playground to show his virtuoso gifts as a theorist but, though many are capable of great complexity of physical behaviour such as ferrites, or photochromics, these still represent a tiny cut through the parameter space of chemistry. Nothing about molecular arrays, little hint of the exhilarating variety of inorganic solids from metal-rich phases to artificial super-lattices.

In writing a guide to the properties of prototype solids in language that chemists whose pleasure is in making new arrangements of matter understand, a different approach will be needed.

J. A. Duffy, Bonding, Energy Levels and Bands in Inorganic Solids, Longmans, London, 1990.

Exotic Properties

There is a saying that 'geography is about maps, and history is about chaps'. As a parallel from the condensed matter sciences we could say that physics looks for 'maps' that summarise and encapsulate the essentials in the behaviour of broad classes of the solid substances that surround us, while chemistry concentrates on the 'chaps', the individual molecules that go to make up a macroscopic ensemble. But we know that history is more than the sum total of individual actions, because people are aware of others when they act, and geography has other dimensions than the abstraction of landscape into diagrams. So it is that a view of physics as based on the behaviour of aggregates, and of chemistry as derived from its microscopic components, is a caricature of the complex happenings in the real world, albeit one sufficiently potent to survive into many of the best known textbooks.

Compare, for example, the approaches of Charles Kittel's 'Solid State Physics', with its sweeping overview of paradigms and Albert Cotton and Geoffrey Wilkinson's 'Inorganic Chemistry', the latter comparable in its original examples and definitions with the Oxford English Dictionary (and with a plot that is in some respects equally hard to discern). This traditional difference of approach, however, is one that has significantly bedevilled the writing of textbooks about what, in its broadest sense, can be called condensed matter (though really in its broadest sense we might have to include biology!). So much is this the case that of the two texts just cited, the former treats no solid with more than about three atoms in each repeat unit while the latter scarcely acknowledges the existence of solids not composed of molecules. The phrase 'solid state chemistry' was therefore, until only a few years ago, close to being an oxymoron.

But now all that has changed and, along with biological chemistry, the chemistry of the solid state is the fastest growing part of the subject. It must be said that the reason why this is so is largely because of interest in exotic physical properties like superconductivity, although the increasing importance of solid-state batteries and catalysts should not be forgotten. Such a realignment of interests within the academic and industrial chemical communities has led to a dearth of books that tread the middle ground between physics and chemistry, which (apart from its numerous other excellent qualities that I shall refer to in a moment) is one reason why Tony Cox's book 'Transition Metal Oxides' was so welcome.

Not so long ago all but a few chemists would have thought metal oxides, including those of the transition metals, as comparatively boring. Yet the range of properties they exhibit (metals, insulators, magnets, superconductors, ferroelectrics) makes them an apt testing ground for the major theories of solid state electronic behaviour. Cox is a sure-footed and convincing guide to the subtleties of these theories, leading the reader from the simple paradigms of ionic and metallic bonding into the labyrinths of electron localisation, variable-range hopping and minimum metallic conductivity. Over the whole hover the spirits of Neville Mott, John Goodenough and Stuart Anderson, each the instigator of seminal insights into the properties of solids, and the latter pair indeed, animators of the Oxford school of solid state chemistry, now alas largely dispersed. Tony Cox's prose is beguilingly fluent, and reminds us that the art of explaining complex matters to a wider audience remains one of the most elusive, as well as rewarding, of the arcane skills of academia.

Transition Metal Oxides: An Introduction, P. A. Cox, Clarendon Press, Oxford, 1992.

Magnetics for Chemists

The title of Olivier Kahn's book 'Molecular Magnetism' begs several questions. Is there such a thing as molecular magnetism, distinguishable from other kinds that have different adjectives attached (such as personal magnetism or animal magnetism)? Is 'magnetism' to be construed as a property of a molecule, or is it inherently a property of an ensemble? The questions are important because the central theme, unstated at the outset but becoming ever more evident as the book proceeds, is to establish principles that will enable chemists (and the book is addressed principally to chemists) to assemble clusters or arrays of molecules whose collective properties are governed by interactions between the constituents and then to rationalise these properties through fundamental theories. The principal interaction mechanism considered is magnetic exchange, of course, and the microscopic mechanisms of super-exchange and double exchange are exhaustively gone into. However, other, less well-known sources of co-operativity also receive attention, such as the way in which the change of metal-ligand bond length accompanying a high-spin to low-spin transition on one molecule transmits its effect to its neighbours.

The starting point, though, is the individual molecule, almost exclusively in the form of the classical Werner-type of coordination complex in which a transition metal or lanthanide ion is embedded in a coating of organic ligands but in which the unpaired d- or f-electrons are only modestly delocalised away from the central metal. The magnetic behavior of crystals made up from such molecules, considered in the first instance as non-interacting, was the stomping ground for armies of theoretically inclined inorganic chemists in the 1960s and 70s, operating under the banner of ligand field theory. One may doubt quite seriously whether any useful purpose is served by going over all this old ground again, and much of the content of the early chapters of Kahn's book is easily accessed through the standard textbooks of Ballhausen and Figgis, not to mention the classic works of Condon and Shortley and of Griffith. The contents of most of the extensive appendixes can also be found in these books. Novelty begins to emerge later in the book, when interactions between metal ions in dimeric or more extended centers are treated. A wealth of formal cases is considered (binuclear, trinuclear-symmetrical and unsymmetrical-isotropic and anisotropic interactions) that will helpfully guide chemists toward rationalising properties of the multifarious compounds turned up adventitiously even by what we sometimes call 'directed' syntheses.

More penetrating in terms of rationalising interaction mechanisms are the descriptions of the various approaches starting from molecular orbital or valence bond limits initiated by Anderson 30 years ago and systematised for the use of solid state chemists by Kanamori and Goodenough. Kahn shows convincingly, and in language familiar to chemists, how what he charmingly labels the 'rustic' approaches to conjugated molecules like the Huckel model fail completely as a starting point for rationalising magnetic interactions. He sets out the formalism of two apparently disparate but in the end equivalent models based on 'natural' and 'orthogonalised' magnetic orbitals and coins some useful phrases such as the 'active electron approximation'. And yet the nagging thought persists that we have been here before, which brings me back to my original question.

The distinguishing features of molecular magnetism as expounded by Kahn are, first, that the materials in question contain only localised magnetic electrons (that is, the book deals only with magnetic insulators) and, second, that for the most part they have quite low point and space group symmetries as a result of elaborate and often quite beautiful molecular architectures brought about by the tailor-made organic ligands. Physicists often use the word 'zoo' in such cases to express their horror at having to search for first principles among such complexity, but in the present case the principles have been available for a long time. Indeed, in his preface Kahn pays tribute to Van Vleck and Anderson, to whose names the others mentioned above should certainly be added. Where then does that leave 'molecular' magnetism? In my view there is a seamless web of increasing structural complexity from binary oxides to metal-organic clusters and multi-metallic enzyme active sites. The central themes of molecular magnetism are then (as I just stated) some beautiful and subtle chemistry, enabling coordination geometry and crystal packing to be tuned to a fine degree, and the fact that these compounds are unusual among magnetic materials in being transparent and colored. Two deficiencies of this book are that it says nothing about the way to actually make molecular magnets or about their electronic excited states. Twenty years ago we found a class of organic-soluble transparent magnets that changed colour at the Curie temperature—what fun! Kahn's book is Cartesian and didactic: the fun (and there is plenty of it) is mostly left to the imagination.

Molecular Magnetism, O. Kahn, VCH, New York, 1993.

A Magnetic History

Magnetism has always been the most mysterious of the natural forces. The word itself, like gravity, long ago entered the English language as metaphor and symbol. But unlike gravity, which is all-pervasive and always attractive, magnetic force appeared doubly mysterious. First, magnetism could repel as well as attract. Second, it was not a universal property of matter; only a few materials, such as lodestone, which was precious for early use in direction-finding, exhibited it. The sheer universality of gravity, by contrast, hid it from scrutiny until long after magnetism's discovery. As a result, the manifestations of magnetic interaction as seen with the naked eye must always have appeared bizarre and, in a certain sense, unnatural.

Thus, magnetism has carried with it a strong sense of the mysterious and even, at certain periods of history, the occult. It could be argued, therefore, that demystifying and eventually controlling and exploiting this remarkable force represents an even greater triumph of human reason and the power of the scientific method than the conquest of gravity by Kepler, Newton and

Einstein. Against this background, Gerrit Verschuur has written a history of magnetism, not just as a major force of nature but also as a paradigm of the whole process of scientific discovery.

The sweep of Verschuur's story is breathtaking: from Pliny the Elder to the Cosmic Background Explorer, via such giants of 19th-century science as Oersted and Ampere, Faraday, Maxwell and Hertz. The path is a well-trodden one from a historical point of view, and in a book whose purpose is to tell a story, perhaps the author can be forgiven for relying nearly exclusively on secondary sources: the most frequently quoted work among the references is the Dictionary of Scientific Biography.

Still, much as the first impulse of a scientist scanning a review article on one's own speciality is to look for one's own name in the references, so I hope that, in view of my affiliation, I can be forgiven for looking especially closely at Verschuur's chapter on Faraday. It begins unpropitiously by ascribing to Humphry Davy the isolation of 47 new elements (probably a slip of the pen, though surprisingly escaping the editor's notice). Seven would be nearer the mark if, in addition to the Group 1 and 2 metals he isolated at the Royal Institution, we include his identification of iodine in Paris in 1813, assisted by Faraday. Also, so far as I am aware, there is no evidence to support Verschuur's claim that Faraday ever worked as an unpaid assistant to Davy before being formally hired.

These are small factual blemishes, but they lower confidence in the accuracy of other facts related. More surprisingly, the most complete biography of Faraday in recent times (*Michael Faraday* by L. Pearce Williams) does not appear in the bibliography. An irritating feature of Verschuur's exposition, which, after all, is gripping enough in its own right, is the interjection of sections of homespun philosophy, such as *obiter dicta* about inspiration in art, science and religion. Furthermore, a trick much used in fiction, but out of place in a historical narrative, exacerbates the irritation: namely, the self-conscious introduction of the narrator's hindsight. This is used in several places to support the author's emphasis on the time dimension, not in a historical sense but as a vital component of the physical phenomena being expounded.

Moving to more modern times, and to the book's conclusion, yet more contentious generalisations crowd in, culminating in a personal philosophy of scientific progress. There is nothing wrong with trying to draw conclusions from history, of course, but history and philosophy remain separate disciplines, even in science. Verschuur's book is subtitled a history; it is a pity that he did not stick more closely to that appellation.

Hidden Attraction: The Mystery and History of Magnetism, Oxford University Press, New York, 1993.

part 5

EPILOGUE

Learning the Rules of the Game

Speech at the 10th anniversary celebration of the Japan Advanced Institute of Science and Technology, Kanazawa, 2001.

Environments for research

In British society, coming-of-age celebrations usually take place on the twenty-first birthday, although for many civil purposes, eighteen is the age of majority. For a human being, ten is quite a young age, not even at the threshold of adolescence. In the world of institutions, however, especially in the fast moving world of the early twenty-first century, ten years is more than enough for a newly founded organisation to develop a personality and make an impact. In that spirit I would like to begin this tribute on the tenth anniversary of the foundation of the Japan Advanced Institute for Science and Technology (JAIST) by congratulating the Institute most warmly on its swift transition from a concept (and later a building site) to the fully fledged institution that we now see, with its own quite distinctive character.

Given the status of JAIST as a graduate research university, this milestone in its history also provides an opportunity to reflect on the place of research in the environment of a university. In particular it provides the backdrop against which we can examine the perennially fascinating question of what is the most effective medium for conducting research in the twenty-first century—a research institute or a university.

In using this occasion to tackle such a question, I must begin with a disclaimer: although the system of research universities is well established in the United Kingdom, I do not wish to go into any detail about how such affairs are organised in my own country. Not that I want to keep anything from you about a system which (in my opinion) has much to commend it. Rather, I would like to concentrate attention on wider and more general issues.

Why do research?

At the outset, an even wider and more general issue is: why are we all so enthusiastic about conducting more and more research? What is the justification for the considerable expenditures that so many nations incur in pursuit of this somewhat esoteric activity? There are two: generating new knowledge about the world we all inhabit is both good in itself, and an essential starting point for new technologies, out of which flow the products and services that underpin and sustain modern civilisation. Research can therefore be characterised on the one hand as a cultural activity, one of those attributes that renders us human. But if that were all, the budgets of our research councils would surely be closer in size to those of our arts councils! For the inescapable truth, known to politicians and industrialists, is that understanding is a necessary prerequisite for controlling and exploiting nature.

The human endeavour of uncovering the inner workings of the world around us is epitomised by the words at the head of this page: learning the rules of the game. The book of nature has a mighty vocabulary, from the outermost universe through the complexities of biology down to the ultimately simple structure of quarks and gluons, but to read it we have to know the grammar. One of the most successful men of all time laying bare the laws of nature was Michael Faraday who, without the benefit of any higher education, from 1812 to 1863 laid the foundations of electrochemistry, electromagnetism and finally field theory by his unremitting work at the Royal Institution in London. Therefore, I can do no better than refer to his words, spoken to an audience of young people at the Christmas Lectures 'On the Various Forces of Nature' that he gave first in 1845. They are quoted earlier in this book (p. 14).

But knowing the grammar is not just satisfying: it is necessary equipment for functioning in the world. Let me give a salutary modern example. Most people will remember the Challenger disaster, when a rocket carrying astronauts blew up after launch from Cape Canavaral. Naturally there was a high level enquiry and, among others, Richard Feynman, one of the twentieth century's greatest physicists, gave evidence. In a simple experiment, performed in front of the investigating panel in a manner worthy of Michael Faraday, by dipping a piece of rubber sealing ring into a glass of iced water, he showed how it lost its elasticity when it was cooled. The catastrophic disaster had been caused by the failure of just such a ring, between two stages of the rocket, brought about by exceptionally cold weather on the morning of the launch. In fact, engineers had cautioned against launching the rocket that morning for this very reason, but the NASA management was anxious to maintain the timetable: budgeting discussions were scheduled to take place very soon and they were keen for a public demonstration.

After the demonstration, hinging on a very simple matter of materials science, Feynman wrote his own report to the U.S. government. Perhaps even more important than identifying the cause of the disaster was the conclusion that he drew: while management, planning and organisational issues are certainly not unimportant, and have their place in preparing for decision making, yet they can never transcend the laws of nature. In a memorable phrase, redolent of the Bible, he concluded that 'nature is not mocked'. The message from this sad episode is clear; we had better, first of all, know the rules by which the game of nature is played and, second, take them fully into account when we decide our courses of action.

Institutes and universities

What I have said so far throws into greater relief the question posed at the outset: what is the best environment for carrying out the programme of identifying and codifying these 'rules of the game'? To answer that question we need to look more closely at the defining characteristics of free-standing research institutes and of universities as platforms for such an endeavour. Immediately it is clear that each one has both advantages and drawbacks in the perennial search for efficiency and economy.

The research institute model

A priori, one might think the answer was clear; a dedicated research institute should be the most efficient way of getting answers to important questions. Consider the procedure: the steps required are self evident:

- Define the problem;
- Assemble the team;
- Apportion the tasks;
- Monitor the progress.

Such an approach also fosters inter-disciplinarity, frequently a fruitful way of approaching complex issues, because the make up of the team is not tied in any way to the traditional disciplinary structure of university departments. It is an approach more familiar in the applied science and commercial environment, where task forces are formed and dissolved to meet problems as they arise. For example, a team containing polymer chemists, polymer physicists and materials scientists could tackle the problem that vexed the NASA engineers. Since so much contemporary research is multidisciplinary, for example chemistry in biotechnology, this appears an ideal approach. Furthermore, the growth in the use of large and expensive research tools, such as neutrons and synchrotron X-rays, necessitates bringing them together in dedicated institutes where users from many smaller laboratories can go to carry out their experiments.

So what are the drawbacks to this apparently effective model? I see two principal ones, the first organisational and the second intellectual. The managerial problem arises when the task for which the institute was established has been accomplished: how do you close it down? People have been hired, buildings built, equipment purchased and so on. There is no easy answer to this problem, and pain and disappointment are almost inevitable consequences. But an even greater danger lurks if a freestanding institute continues in business unchanged for too long, that of fossilisation. Research programmes can easily become self-justifying and teams of scientists, like any other profession, complacent as they grow old together. Is there a better model?

The university model

Conversely, the strengths of the university as a platform for advanced research address both of the above weaknesses. Unlike a research institute, a university is constantly renewing itself. Oxford has had a university for nigh on a thousand years, but the university of the twenty-first century bears absolutely no resemblance to that of the twelfth. I taught chemistry for many years in a room that had been built before the subject was put on its earliest secure footing by Lavoisier at the end of the eighteenth century! One is reminded of the Greek philosopher Heraclitus, who noted that when we look at a stream, we never see the same water twice, yet the stream remains. What constitutes this endless novelty in a university? The students. It is the continuous flux of young people, year on year, that guards against fossilisation, not only in their ever changing personnel, but in the requirement that places on their seniors (whose lifetime in the institution greatly exceeds the three years or so of student life) to explain and justify, to re-examine old beliefs and admit new hypotheses.

Of course the university system is not without its drawbacks. Principal among these is the traditional disciplinary structure imposed by undergraduate teaching, which certainly does not map on to the majority of present day topics for research. But even that provides us with a guarantee that those undertaking the research have a thorough preparation in the techniques of their discipline. After all, one cannot be usefully interdisciplinary except from a solid disciplinary basis. The teams or task forces that I alluded to earlier comprise people whose contribution is their disciplinary professional expertise.

A second constraint on the university system in efficiently prosecuting research is that the efforts of faculty must also be directed towards teaching, thus eating into the time and energy available for the pursuit of new knowledge. Yet such a division of attention also has its positive side, for it is precisely through interaction with students that the latter are brought to the boundaries of current knowledge and, by experiencing themselves the enthusiasm and dedication of their teachers, are brought into direct contact with the research enterprise. An additional constraint (probably not applying too seriously to JAIST) is the necessary division of budgets between research and teaching.

Finally, let it also be acknowledged that teaching, it its broadest sense, is an essential element in the proper conduct of research. Not only is this the case because techniques and procedures have to be transmitted. Even more important (indeed crucial to sustain innovation) is the issue of attitude in approaching the unknown. Here we see the fundamental difference between experiments performed as part of laboratory instruction in an undergraduate course, and those carried out by a researcher. In the former case, not only is the answer already known, but the conditions of the experiment have been optimised by the instructors to maximise the chance that correctly applied techniques will produce it. In the research laboratory, not only are the optimum conditions not known, but the question posed may not even have a definitive answer. What teaching at the graduate research level can instil is, above all, a skill in asking the right questions in the chosen field, and an ability to assess the answers in a sceptical yet constructive way. Beginners in research need to know that nature does not yield up its secrets easily; that equipment frequently yields puzzling data; that experiments go wrong, but above all, that the unexpected may just possibly signal a breakthrough into new territory. That is the most exciting outcome of all.

Conclusion

Everything that has been said so far points inexorably to the university as the optimum venue and platform for creative and innovative research. Where the problem can be clearly defined, and the answers urgently awaited, the task force or research team approach has much to commend it, and where longer term programmes or large scale instrumental investments are needed, a dedicated institute may be appropriate. Nevertheless, the uniquely positive feature distinguishing the university is the infinite and continuously changing cohort of young people that it brings to the research adventure. It is through fresh approaches that research flourishes most vigorously, and the new generations are its guarantors.

It will not have escaped notice that the model being commended bears more than a passing resemblance to JAIST. In closing, may I therefore take the opportunity of wishing this fine and almost unique institution every success in the coming century as it takes its due place among the world's leading research universities.

part 6

BIBLIOGRAPHY

It will not have escaped the reader's notice that nowhere in this book are any references to be found. That is because, as explained in the introduction, the text is derived in large part from a series of articles published in a variety of magazines, journals, reviews and so on, each of which had a different policy with regard to referencing source material. In some cases the references are extensive, in others exiguous. Furthermore, because this compilation is intended for the more general reader, it seemed preferable not to clutter it up with scholarly apparatus. That is not to say, though, that justification will not be found in the primary literature for the historical and scientific facts given. (As far as expressions of personal opinion and evaluation are concerned, that of course is another matter.) Interested readers looking for more information are directed to the original articles and the references they contain. For convenience (as well as by way of acknowledging my sources) I give below the articles from which this collection has been compiled.

Part 1: Temples of Science

1. Many Happy Returns at the RI, *Chemistry in Britain*, 30–35 (April 1999).
2. Creating and Communicating Science: The Experience of the Royal Institution, 9th Blackett Memorial Lecture, *Proc. Indian Natn. Sci. Acad.*, **60**, A, 607–617 (1994).

3. The Philosopher's Tree: Faraday Today at the Royal Institution, *Proc. Roy. Inst.*, **70**, 1–20 (1999).
4. Friday Evenings at the Royal Institution, *Proc. Roy. Inst.*, **66**, v–vi (1995).
5. The Royal Institution in Japan, *Sci. in Parliament*, **51**, 34 (1994).
6. The Institut Laue-Langevin: Crucible of European Sciences, *Proc. Roy. Inst.*, **64**, 51–68 (1993).
7. Conversation Rooms, previously unpublished.

Part 2: Some Past Masters

1. 'Mr Secretary, Colonel, Admiral, Philosopher Thompson': The European Odyssey of Count Rumford, *European Review*, **3**, 103–111 (1995).
2. The Root of all Research, *Chemistry in Britain*, 471–473 (June 1995).
3. Michael Faraday as a Materials Scientist, *Materials World*, 374–376 (1995).

Part 3: Some Folks You Meet

1. Christian Klixbull Jorgensen (1931–2001): Inorganic Spectroscopist Extraordinaire, *Coord. Chem. Rev.*, **238–239**, 3–8 (2003).
2. Whereof Man Cannot Speak: Some Scientific Vocabulary of Michael Faraday and Klixbull Jorgensen, *Structure and Bonding*, **106**, 7–18 (2004).
3. Olivier Kahn (1943–1999), *Nature*, **403**, 498 (2000).
4. Molecules and Magnets: Joining Chemistry with Physics. The Legacy of Olivier Kahn, *C. R. Acad. Sci. Paris, Chimie*, **4**, 75–78 (2001).
5. Lord Dainton, St. John's College, Oxford, Notes, 60–63 (1998).

Part 4: Molecules, Solids and Properties

1. Molecular Magnets: The Prehistory, *Notes Rec. Roy. Soc.*, **56**, 95–103 (2002).
2. The Chemistry of Magnets, *Science*, **261**, 431–432 (1993).
3. Magnets Without Metals, *Nature*, **363**, 113–114 (1993).
4. Mixed-Valence Compounds, *Endeavour*, **29**, 45–49 (1970).
5. Superconductors: Past, Present and Future, *Proc. Roy. Inst.*, **65**, 29–46 (1994).
6. Room at the Bottom, *Chemistry in Britain*, 29–31 (July 1996).

7. Molecular Information Processing: Will it Happen? *Proc. Roy. Inst.*, **69**, 85–106 (1998).
8. Synthesis and Properties of Low-Dimensional Compounds, *Endeavour*, **4**, 127 (1980).
9. The Ties that Bind So Many Compounds, *New Scientist*, 50 (10 November 1990).
10. Exotic Properties, *Times Higher Educ. Suppl.* (9 April 1993).
11. Magnetics for Chemists, *Science*, **266**, 145–146 (1994).
12. Hidden Attraction: The Mystery and History of Magnetism, *Physics Today*, **47**, 64–65 (1994).

Part 5: Epilogue

1. Learning the Rules of the Game, unpublished.

Index